Medium Voltage Cable Accessories

A Theoretical & Practical Appraisal

Dr. Derek Goulsbra
CEng. MIET

Nexans

Published by Nexans Power Accessories (UK) Ltd

Unit 2
Aire & Calder Industrial Park
Lock Lane
Castleford
West Yorkshire
United Kingdom WF10 2JA

www.nexans.co.uk

+44 (0) 1977 669966
sales@nexansdirect.com

First published 2012

ISBN: 978-0-9573785-0-6

Foreword

This book is a follow up to an earlier edition entitled "Some thoughts on MV cable accessories". Greater emphasis is placed on modes of failure and ways to overcome potential problems encountered in the field. It is the aim of the book to provide as much information as possible to engineers and jointers who are responsible for specifying and installing these products. It is hoped that by following the information presented, the reliability of accessories will be improved thus saving considerably on the cost of failures due to loss of supply, labour and equipment replacement.

Cable accessories account for the least capital expenditure in the distribution network when compared with switchgear, transformers and cables, but can prove to be the weakest part of it because they have to be assembled on site. This book will describe in detail the workings of accessories and show why common faults can occur and the steps to be taken to prevent problems.

Some of the mathematics and associated assumptions relating to voltage distribution in this work are simplified and two-dimensional. In reality we a looking at complex three-dimensional models, but it is felt that it is unnecessary to proceed down this path in order to explain in broad terms how certain phenomena occur. The author makes no apology for this, but begs the reader understands the purpose of the work.

Nexans Power Accessories (UK) Ltd

Acknowledgements

I am indebted to the help and encouragement given by Gary Halton, Mick Dawn, Emma Hutton and Jon Rhodes of Nexans Power Accessories (UK) Ltd. The following companies have been most helpful in providing information and illustrations and their cooperation is much appreciated.

- CMP Products – www.cmp-products.co.uk
- Ellis Patents Ltd – www.ellispatents.co.uk
- W T Henley Limited – www.wt-henley.com
- Nexans Power Accessories (UK) Ltd – www.nexans.co.uk

Contents

Chapter 1

Introduction

1

The cables used for electrical distribution until the 1960's almost exclusively used impregnated paper to insulate the conductors and were usually three core. A change to polymeric insulation then gradually gathered momentum and with it a change in the technology required for cable accessories, although, ironically some of the first dry terminations were developed for paper cable.

Early so-called dry type accessories for polymeric cables were usually of the push-on variety produced from elastomers and used on single core cable. The sound basic technology employed in the early days ensured that such products continue to be supplied today.

Heat shrinkable terminations were introduced in the late 1960's and had the distinct advantage of being range taking and capable of use on both single and three core cables. Heat shrinkable joints were developed a few years later following the success of terminations and their suitability for both paper and polymeric cables resulted in a major revolution in cable accessory technology. This technology was a complete departure from the previous techniques, which required a cast iron box or lead sheath filled with oil or compound.

Cold shrink products made an appearance a little later and are now proving popular within the industry. The debate as to which system is preferable continues. It is fair to say that the heat shrinkable system is the most versatile and well known, but on single core polymeric cables in particular cold shrink systems are seen to have certain advantages. It is not the intention of the author to express a preference for any one system.

The use of screened elbows to terminate cables is now widely accepted and it will be shown later in the book that this system is theoretically superior to other systems for certain applications.

The early development of dry cable terminations concentrated on three core paper cables in the United Kingdom and single core polymeric cables in the United States of America. The accessories for paper cable were originally developed for paper belted cable rated at 11 kV using heat shrink technology; whilst for polymeric cable elastomeric push on systems were developed principally for 8, 15 and 25 kV.

As with many totally new concepts, it was shown that extensive laboratory development and testing is no substitute for field experience and many of the early terminations failed. The failures were generally not the result of poor installation, as each one was carefully monitored at this stage. The first heat shrinkable terminations developed for paper cable were pole top mounted and therefore exposed to the elements. The materials used were not themselves to blame and the formulation has changed little since that time. Several weak links in the chain were, however, present simple things such as inspection windows in cable lugs allowing water into the termination and the sealant used did not always exclude moisture. However, the major problem was the regular occurrence of breakdown between one core and the lead sheath.

Gradually, as the problems were identified, the weaknesses were eliminated and the products became reliable. At the present time dry type accessories, if installed correctly, will give satisfactory operation for the life of the system. Many terminations installed thirty years ago are still in service today.

As the practices of utilities have change and many of the jointers and engineers have retired, the need to make the installation of accessories less skill sensitive is the main driving force for the manufacturer. However, there are still some fundamental rules which should be obeyed to ensure satisfactory installation and hence trouble free service life.

A constant problem within the industry is the shortage of fully trained jointers to install the ever increasing variety of products in the market place.

The following chapters will examine how accessories work, why they can fail and good practices required ensuring trouble free service. It is hoped that there will be information of interest to engineers and jointers alike which will enable them to better understand the workings of cable accessories and hence reduce the potential for problems at a later stage.

Chapter 2

Electrical breakdown of air and solid dielectrics

In order to understand accessories and how they perform it is necessary to look at the fundamental electrical breakdown of air and solid dielectrics, which are, of necessity, mixed in dry cable accessories.

2.1. Electrical Breakdown of Air

Air at normal temperature and pressure, which means a temperature of 20 °C and pressure of 760 mm Hg, has an intrinsic breakdown strength of 3 MV m^{-1}. This means that if a voltage is applied between two electrodes, which provide a uniform field, and the electrodes are spaced 10 mm apart, electrical breakdown of the air in the form of an arc will occur at 30 kV. This value is the same for a.c. (the peak value), d.c. and impulse. The electric strength is simply found by dividing the applied voltage at breakdown by the electrode separation. In the present example:

Breakdown strength E = Voltage applied / Electrode separation
= 30 kV per cm
= 3 kV per mm
= 3 MV per m

A uniform field is shown in fig.1. By definition lines of equal voltage are spaced equally across the shortest distance of the electrodes.

In practice, the intrinsic electric strength of air is of little use because conductors at high voltage are not conducive to producing a uniform field and the earth plane is often a flat sheet interspersed with bolt heads and nuts for various fixings. At the other end of the scale from a uniform field is the point-plane configuration shown in fig. 2.

In this situation, it will be noted that the equipotential field lines are much closer together near the point electrode and spaced further apart as the earth electrode is approached. If the separation of the electrodes is again 10 mm, it will be found that

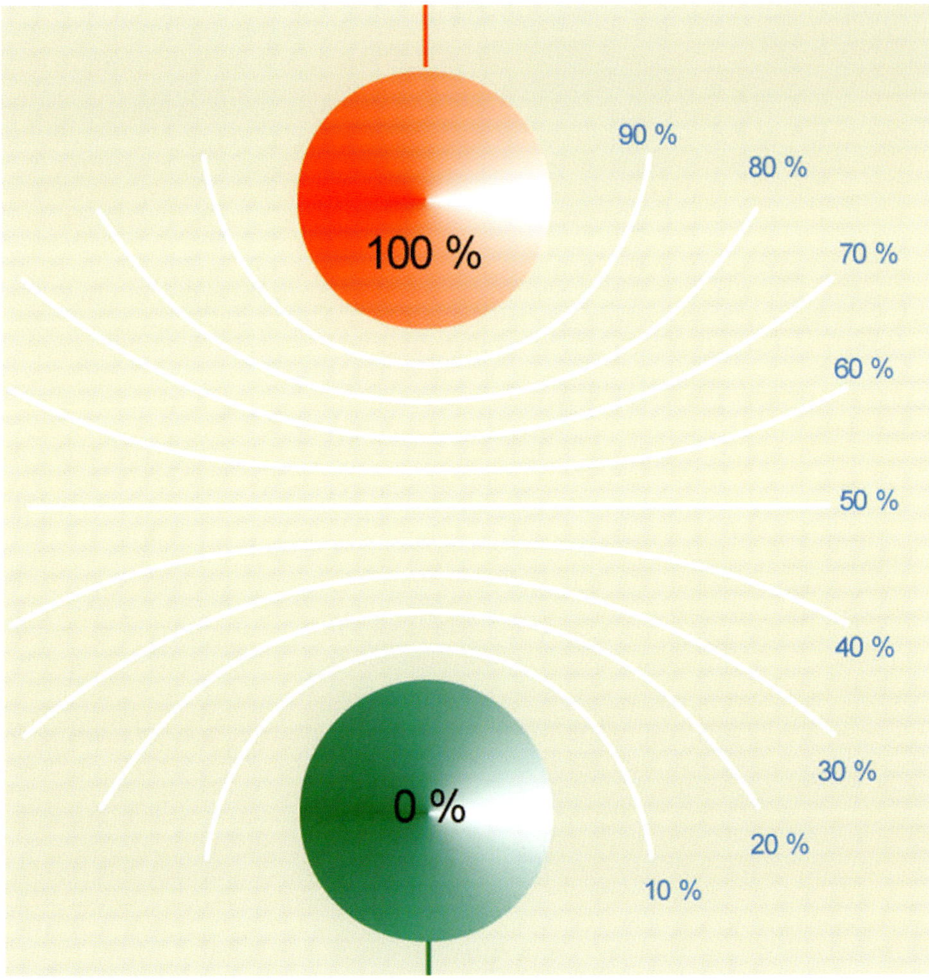

Figure 1
A uniform field

the voltage required to break down the air will be as low as 10 kV, depending upon the sharpness of the point. The reason for this is that as the field lines are close together near the point, the breakdown strength of the air, i.e. 3 MV m^{-1} will be reached with only a few kV applied. At this point the air breaks down locally to create corona around the needle electrode. There will not be complete breakdown of the gap initially because the electric stress nearer the earth plane is very much lower. The corona produced does, however, ionise the air thus reducing its electric strength and if the voltage is increased further, total breakdown occurs at a much lower level than if the electrodes presented a uniform field.

Corona occurs in many situations without causing complete failure of the system – e.g. on overhead power lines and around insulators on these lines, especially when there is rain or fog in the atmosphere.

Whenever air is used as part of the insulation in electrical equipment, there is the possibility of corona occurring. It is not desirable for this to happen in an enclosed space and thus designers ensure that sharp profiles are eliminated and spacing is adequate to prevent corona at working voltage.

It should be borne in mind that corona in an enclosed piece of switchgear or cable box can, if the humidity is high, create various acids, which will eventually corrode the metal work of the equipment and this must be avoided. It will be shown later that this effect can also be caused by badly installed cable terminations.

The above two examples of electric breakdown of air give an insight into what can happen if there are sharp edges or insufficient clearance between high voltage and earth. One final point, if a complete breakdown of air occurs, or there is corona, when the reason for the fault is removed, the air recovers its properties quickly and the system can be re-energised. If, however, the insulation is not air, but a solid, catastrophic failure occurs and this will now be discussed.

2.2. Electrical Breakdown of Solid Insulating Materials

Consider once more the uniform field shown in fig. 1. If the electrodes were completely surrounded by a solid, the field lines would be the same, but the voltage required to produce electrical failure would be higher. Once breakdown occurs, the damage is irreparable as the breakdown path destroys the original material and produces a conducting path.

Table 1. gives some typical values of the breakdown strength of materials widely used in the past and the present day. (Note that these values can vary considerably depending upon test method and sample thickness.)

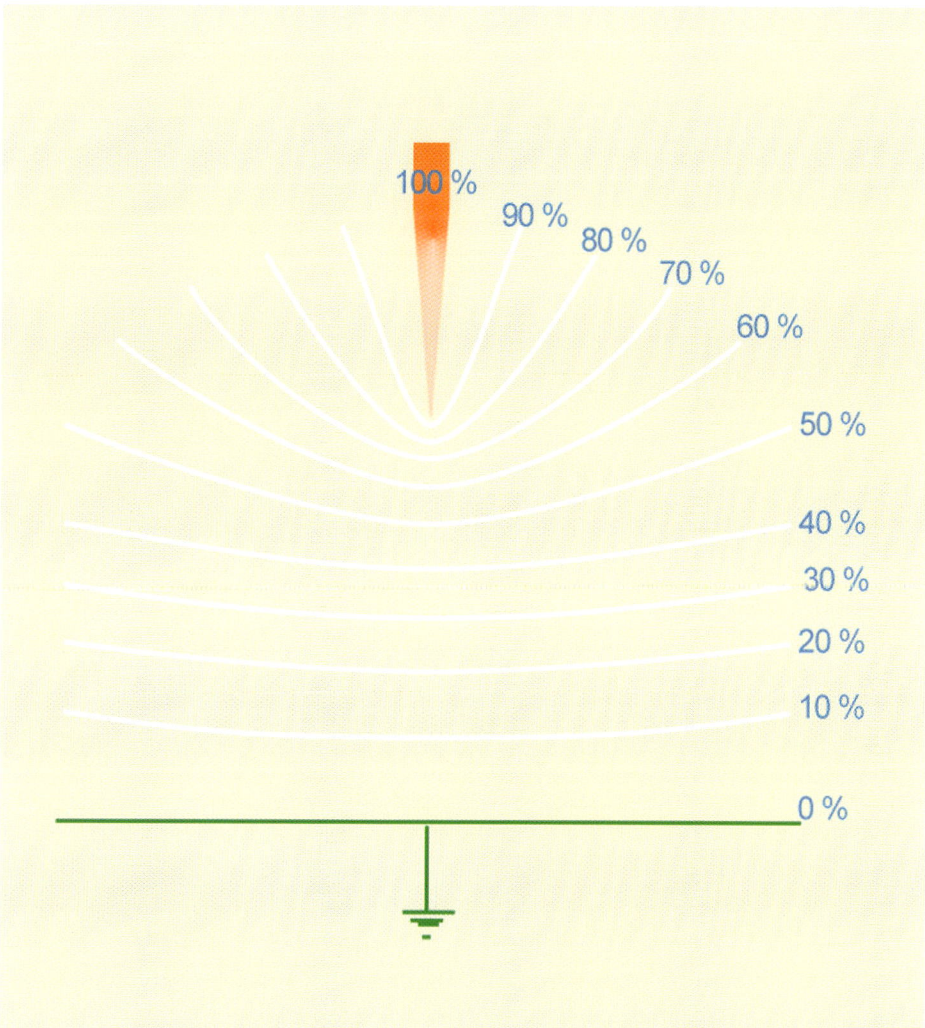

Figure 2
A non-uniform field

Insulating Material	Electric Strength MV m^{-1}
Air	3
Compound (used in old cable boxes)	10
MIND paper (used on cables)	12
PVC (used on LV cables, busbar sleeving etc)	20
XLPE (used on MV & HV cables)	15
Epoxy Resin (used for bushings and dry transformers)	25
Polyurethane Resin (used in cable joints)	15
EVA Tubing (used on terminations and busbar sleeving)	15
Silicone Rubber (used on cable accessories)	20
EPDM (used on push-on including screened elbows)	15

Table 1
Some typical electric strengths of insulating materials

A quick glance at the table may suggest that the higher the electric strength the more appropriate the material. In fact this is not necessarily the case.

Air is used because it is cheap, but air alone requires large separations between live conductors and earth and this leads to bulky and more expensive equipment.

PVC has a high dielectric strength, but if subjected to surface electric stress will fail by tracking in a very short space of time. Further, if it catches fire during a fault it gives off copious amounts of smoke and toxic gases. It has been used extensively on LV cables because it is a relatively cheap material and as it is not highly electrically stressed it performs perfectly satisfactorily. However, it cannot be operated above 70 °C without long term degradation taking place.

Most medium voltage cables now incorporate XLPE as the primary insulation and the fact that it is cross linked means that it can operate at a higher temperature than PVC, typically 90 °C, thus allowing greater current to flow for a given cross section of cable. It is more expensive than PVC and has a lower electric strength, but its other properties make it far more suitable for this application.

Epoxy resin has a very high electric strength, but its uses are restricted to such applications as switchgear and transformer bushings, which are factory cast, and the insulation for dry transformers. It can be used in cable joints, but the cost is much higher than the generally accepted polyurethanes.

EVA is the base material used in most heat shrink formulations and has the advantage, unlike most of the materials mentioned above, that it can be produced in a heat shrinkable form with a shrink ratio of at least 3:1.

Silicones are gaining widespread use in the insulation field as the price comes down and they are often used in cold applied cable accessories because they can be stored on a holdout prior to installation. They are also suitable for push-on technology including screened elbows.

Let us now consider the breakdown of a solid in a non-uniform field situation similar to that of fig. 2. If the voltage applied to the needle electrode should exceed the breakdown strength of the material, the solid will irreparably break down in the region of the needle. This breakdown will cause a small channel, which effectively extends the point of the electrode, so reducing the distance to earth and further increasing the electric stress across the insulation. The long term effect is that a series of channels are formed which eventually cross to the earth plane, resulting in complete breakdown. This phenomenon is called treeing and is shown in fig.3. The problem with this type of breakdown is that it can take years to reach final failure and is not shown up in commissioning tests.

Having looked in detail at the breakdown of air and solids separately, an examination of the situation when air and solids are mixed will be carried out.

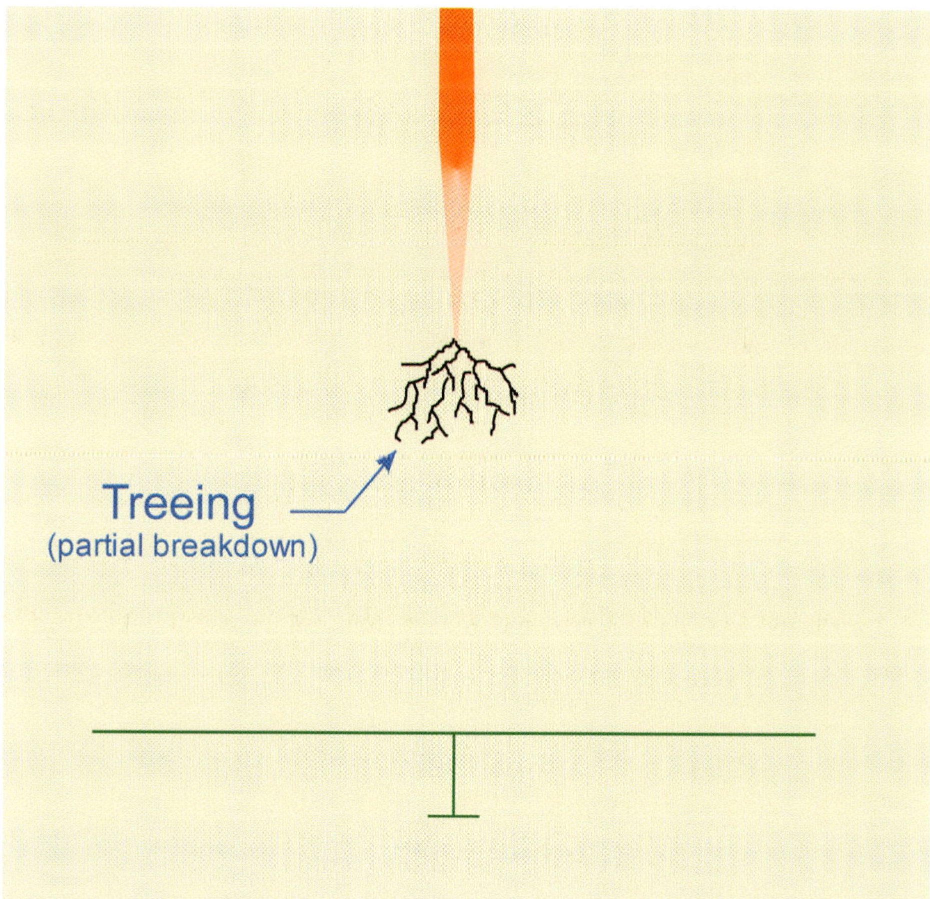

Figure 3
Treeing in solid dielectric in non-uniform field

2.3. Effect of air/solid combinations on voltage distribution

When air and solid insulation combine to provide the total path between two conductors at different potentials the voltage distribution is very different to the situation when only one dielectric is present. This is because the different dielectrics act as a voltage divider in much the same way as resistors with different values in series do. In this case we have a capacitive divider. Figure 4 shows a simple case where air is trapped in a solid dielectric. This is not a common occurrence now, but was a problem in the past when manufacturing techniques were not guaranteed to provide perfect castings, mouldings and extrusions on a consistent basis.

The example shown in fig. 4 will be addressed in some detail. In practice the model and hence the mathematics would be more complex, but it does show an uneven sharing of voltage. Thus, whilst the applied voltage is well below the breakdown strength of the insulation, the stress across the void is high enough to cause breakdown of the void. The result of this is that the discharges eventually break down the solid insulation and treeing can again result as shown in fig. 3 with ultimately complete breakdown. To understand this in more detail the equivalent circuit of fig. 4 is taken a stage further in fig. 5.

2.3.1. Voltage distribution between infinite parallel plates

The voltage across the void, V_2, is calculated as follows:

$$V_2 = IX_{C2}$$

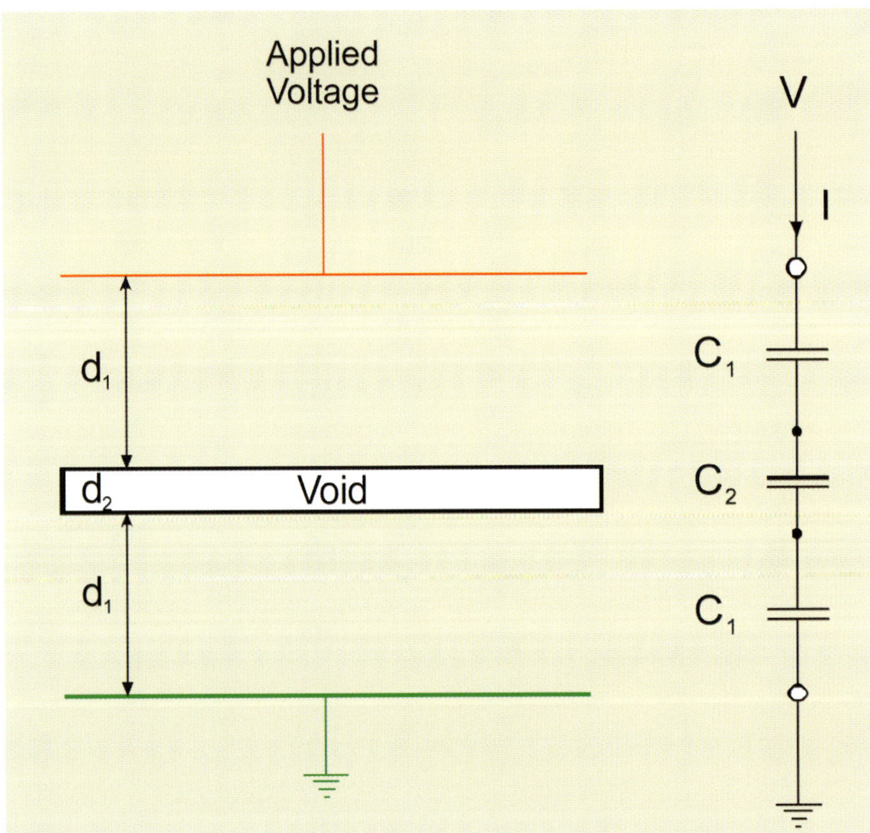

Figure 4
Equivalent circuit for a void entrapped in a solid dielectric

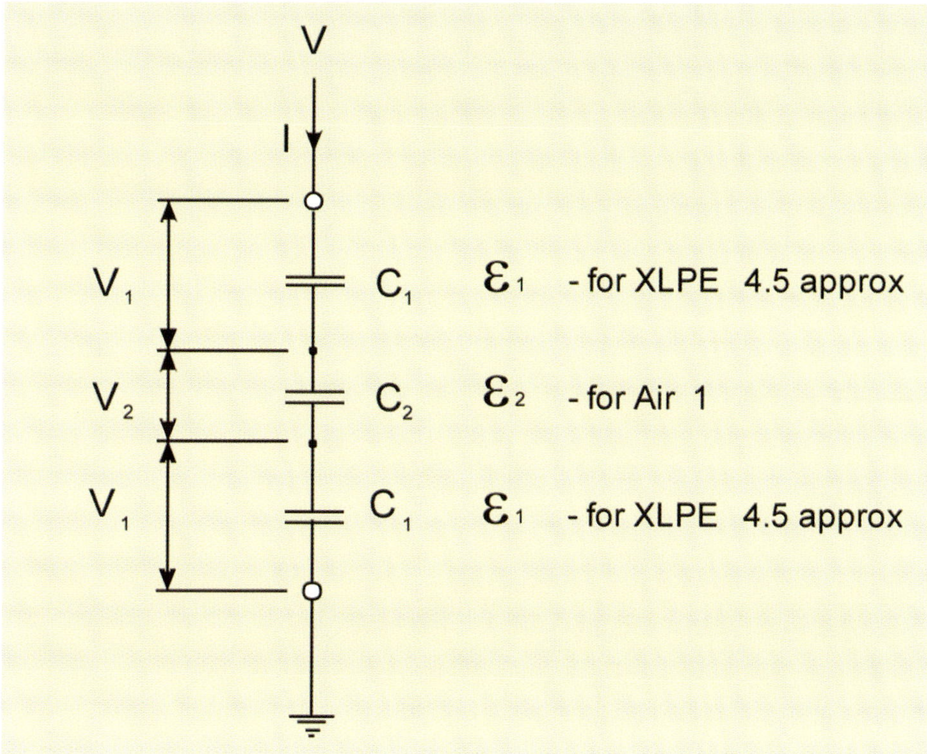

Figure 5
Voltage divider effect as a result of a void

where

V_2 is the voltage across the void
I is the capacitive current flowing between electrodes
X_{C2} is the capacitive reactance of the void

now,

$$X_{C2} = \frac{1}{2\pi f C_2}$$

where

f is the supply frequency (50 Hz)
C_2 is the capacitance of the void

now,

$$C_2 = \frac{\varepsilon_0 \varepsilon_1 A_2}{d_2}$$

where

ε_0 is the permittivity of free space (constant)
ε_1 is the permittivity of the dielectric, which in this case is 1
A_2 is the cross sectional area of the void
d_2 is the depth of the void

Substituting the various equations above gives the voltage across the void as:

$$V_2 = I\frac{d_2}{2\pi f\varepsilon_0\varepsilon_2 A_2}$$

$$\text{i.e. } V_2 \sim \frac{1}{\varepsilon_2}$$

In other words, the voltage drop is proportional to the inverse of the relative permittivity.

It can be similarly shown that:

$$V_1 = I\frac{d_1}{2\pi f\varepsilon_0\varepsilon_1 A_1}$$

$$\text{i.e. } V_1 \sim \frac{1}{\varepsilon_1}$$

It should now be clear that as the permittivity of air is 1.0 and for solid dielectrics, considerably higher, then a disproportionate amount of voltage appears across the void.

Let us assume that $A_1 = A_2 = A$

Then we can determine the voltage across the void in terms of the applied voltage, permittivities, thickness of insulation and depth of void as shown below:

$$V = \frac{I}{2\pi f\varepsilon_0 A}\left[2\left[\frac{d_1}{\varepsilon_1}\right]+\left[\frac{d_2}{\varepsilon_2}\right]\right]$$

$$V = \frac{I}{2\pi f\varepsilon_0 A}\left[\frac{2d_1\varepsilon_2 + d_2\varepsilon_1}{\varepsilon_1\varepsilon_2}\right]$$

$$I = V\frac{2\pi f\varepsilon_0 A\varepsilon_1\varepsilon_2}{2d_1\varepsilon_2 + d_2\varepsilon_1}$$

now,

$$V_2 = IX_{C2}$$

therefore,

$$V_2 = V\left[\frac{2\pi f\varepsilon_0 A\varepsilon_1\varepsilon_2}{2d_1\varepsilon_2 + d_2\varepsilon_1}\right]\left[\frac{d_2}{2\pi f\varepsilon_0\varepsilon_2 A}\right]$$

$$V_2 = \frac{d_2\varepsilon_1 V}{2d_1\varepsilon_2 + d_2\varepsilon_1}$$

If we put some values to the above, it will give a clear indication of how a void can distort the electric field and voltage sharing.

$V = 19$ kV
$d_1 = 4$ mm
$d_2 = 1$ mm
$\varepsilon_1 = 4.0$
$\varepsilon_2 = 1.0$

This gives a value for the voltage, V_2, across the void as 6.33 kV rms, which has a peak value of 8.95 kV. This means the stress across the void is 8.95 MV/m, which is almost three times the breakdown strength of air.

The above values were chosen as the phase voltage of a 33 kV system, i.e. 19 kV, and the total thickness of cable insulation on a 33 kV cable, i.e. 8 mm. The depth of the void is chosen arbitrarily, but would equate to the thickness of the extruded earth screen. Whilst this is a two dimensional linear field it gives a clear indication of the uneven voltage distribution of mixed dielectrics. A more realistic example pertaining to a cable is now considered in the following sections

2.3.2. Voltage distribution between concentric cylinders

Before considering the voltage distribution across mixed cylindrical dielectrics a few basic equations will be developed for the capacitance and electric stress between two cylindrical electrodes as shown in fig. 6 in order to aid the understanding of subsequent workings.

The electric flux density per unit length at radius r is given by

$$D = \frac{\Psi}{A} = \frac{Q}{2\pi r}$$

Where

D is the flux density
Ψ is the total flux across the dielectric
A is the area at radius r
Q is the charge

The above units are expressed per unit length.
By definition, the electric stress, E, at radius r is

$$E = \frac{D}{\varepsilon_0 \varepsilon_1} = \frac{Q}{2\pi \varepsilon_0 \varepsilon_1 r} \qquad (1)$$

where

ε_0 is a constant
ε_1 is the permittivity of the dielectric between electrodes

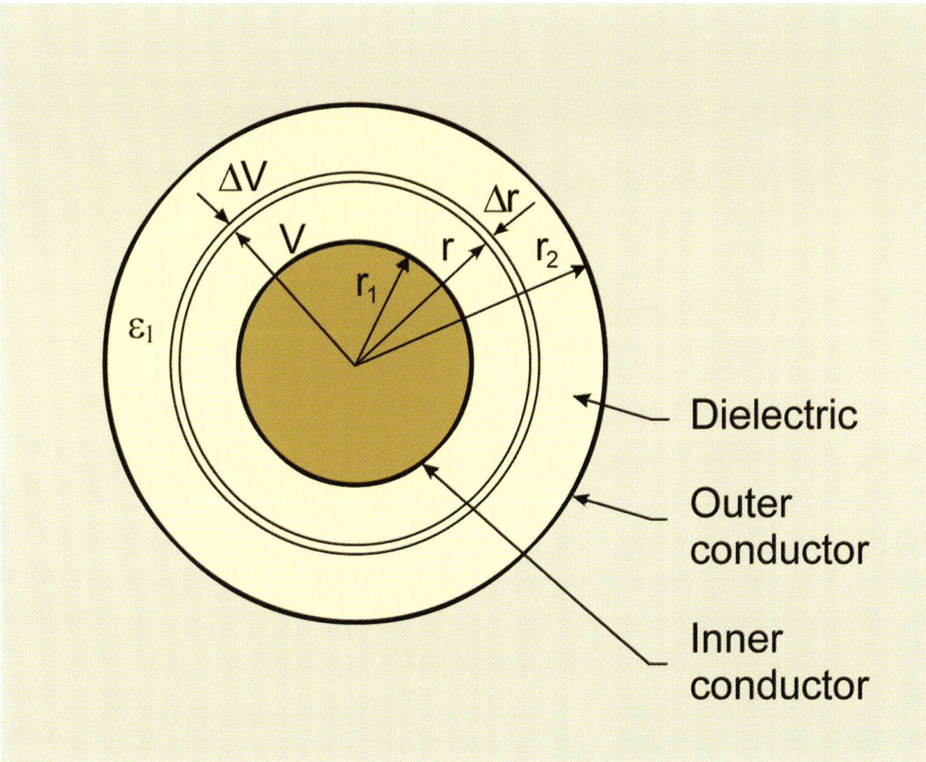

Figure 6
A cylindrical model

If the voltage across the element is ΔV, then the electric stress at r is given by:

$$E = \frac{V - (V + \Delta V)}{\Delta r} = -\frac{\Delta V}{\Delta r}$$

$$\Delta V = -E\Delta r = -\frac{Q}{2\pi\varepsilon_0\varepsilon_1 r}\Delta r$$

The negative sign is because the voltage is decreasing as r increases
The voltage between electrodes, V, is found as follows:

$$V = -\frac{Q}{2\pi\varepsilon_0\varepsilon_1}\int_{r_2}^{r_1}\frac{dr}{r}$$

$$V = \frac{Q}{2\pi\varepsilon_0\varepsilon_1}\ln\left[\frac{r_2}{r_1}\right]$$

now, capacitance per unit length is given by:

$$C = \frac{\text{Charge}}{\text{Potential Difference}} = \frac{Q}{V} = \frac{2\pi\varepsilon_0\varepsilon_1}{\ln\left[\frac{r_2}{r_1}\right]}\ \text{Farads / metre} \quad (2)$$

hence, from (2),

$$\frac{Q}{2\pi\varepsilon_0\varepsilon_1} = \frac{V}{\ln\left[\frac{r_2}{r_1}\right]}$$

substituting into equation (1) we have

$$E = \frac{V}{r\ln\left[\frac{r_2}{r_1}\right]}$$

2.3.3. **Voltage distribution between concentric cylinders with two dielectrics**

Using the equations developed in the above section, we can now proceed to look at two dielectrics between cylindrical electrodes as shown in fig. 7.

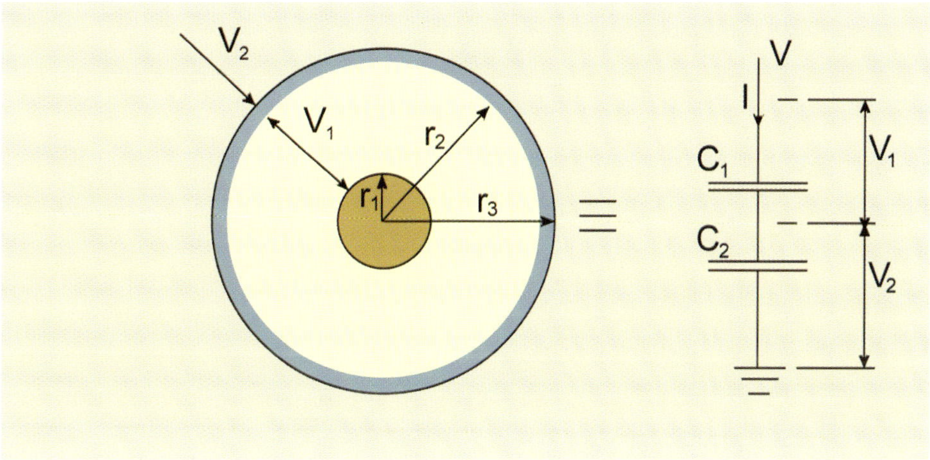

Figure 7
Cylindrical mixed dielectric model

The voltage across the outer, air layer, V_2, is calculated in a similar manner to that shown in the previous example of parallel plates as follows:

$$V_2 = IX_{C2}$$

where

V_2 is the voltage across the air
I is the capacitive current flowing between electrodes
X_{C2} is the capacitive reactance of the air

now,

$$X_{C2} = \frac{1}{2\pi fC_2}$$

where

is the supply frequency (50Hz)
C is the capacitance of the air

now, for a cylindrical electrode we have shown that

$$C_2 = \frac{2\pi\varepsilon_0\varepsilon_2 L}{\ln\left[\frac{r_3}{r_2}\right]} \qquad (3)$$

where

ε_0 is permittivity of free space (a constant)
ε_2 is the relative permittivity of the dielectric, which in this case is 1 (for air)
L is the length of the air cylinder
r_2 is the inner radius of the layer of air
r_3 is the outer radius of the layer of air (i.e. earth)

thus

$$X_{C2} = \frac{\ln\left[\frac{r_3}{r_2}\right]}{4\pi^2 f\varepsilon_0\varepsilon_2 L} \qquad (4)$$

similarly,

$$X_{C1} = \frac{1}{2\pi f C_1}$$

and

$$C_1 = \frac{2\pi\varepsilon_0\varepsilon_1 L}{\ln\left[\frac{r_2}{r_1}\right]}$$

where

ε_0 is permittivity of free space (a constant)
ε_1 is the relative permittivity of the solid dielectric
L is the length of the dielectric cylinder
r_1 is the inner diameter of the dielectric (dia. of conductor)
r_2 is the outer diameter of the dielectic

thus

$$X_{C1} = \frac{\ln\left[\frac{r_2}{r_1}\right]}{4\pi^2 f\varepsilon_0\varepsilon_1 L}$$

now,

$$V = V_1 + V_2$$

$$V = I(X_{C1} + X_{C2})$$

Using the above equations gives

$$V = \frac{I}{4\pi^2 fL\varepsilon_0}\left[\frac{\ln\left[\frac{r_2}{r_1}\right]}{\varepsilon_1} + \frac{\ln\left(\frac{r_3}{r_2}\right)}{\varepsilon_2}\right]$$

$$V = \frac{I}{4\pi^2 fL\varepsilon_0}\left[\frac{\varepsilon_2 \ln\left(\frac{r_2}{r_1}\right) + \varepsilon_1 \ln\left(\frac{r_3}{r_2}\right)}{\varepsilon_1\varepsilon_2}\right]$$

therefore

$$I = \frac{4\pi^2 fL\varepsilon_0\varepsilon_1\varepsilon_2 V}{\varepsilon_2 \ln\left(\frac{r_2}{r_1}\right) + \varepsilon_1 \ln\left(\frac{r_3}{r_2}\right)} \qquad (5)$$

now

$$V_2 = IX_{C2} \qquad (6)$$

combining equations 4, 5 & 6

$$V_2 = \frac{4\pi^2 fL\varepsilon_0\varepsilon_1\varepsilon_2 V}{\varepsilon_2 \ln\left(\frac{r_2}{r_1}\right) + \varepsilon_1 \ln\left(\frac{r_3}{r_2}\right)} \cdot \frac{\ln\left(\frac{r_3}{r_2}\right)}{4\pi^2 fL\varepsilon_0\varepsilon_2}$$

$$V_2 = \frac{V\varepsilon_1 \ln\left(\frac{r_3}{r_2}\right)}{\varepsilon_2 \ln\left(\frac{r_2}{r_1}\right) + \varepsilon_1 \ln\left(\frac{r_3}{r_2}\right)}$$

Solving for V_1 and V_2 using the equation developed earlier for voltage at radius r allows the voltage distribution to be plotted from the conductor to earth and this is shown in fig. 8.

Note that the electric stress is calculated at any point as the slope of the voltage curve and it is seen that the stress is highest at the solid/air interface.

Applying values to the above equation once more shows an uneven voltage distribution between the conductor and earth. Taking values for a 95 mm² 33 kV cable,

$V = 19$ kV
$r_1 = 7$ mm
$r_2 = 15$ mm
$r_3 = 16$ mm
$\varepsilon_1 = 4.0$
$\varepsilon_2 = 1.0$

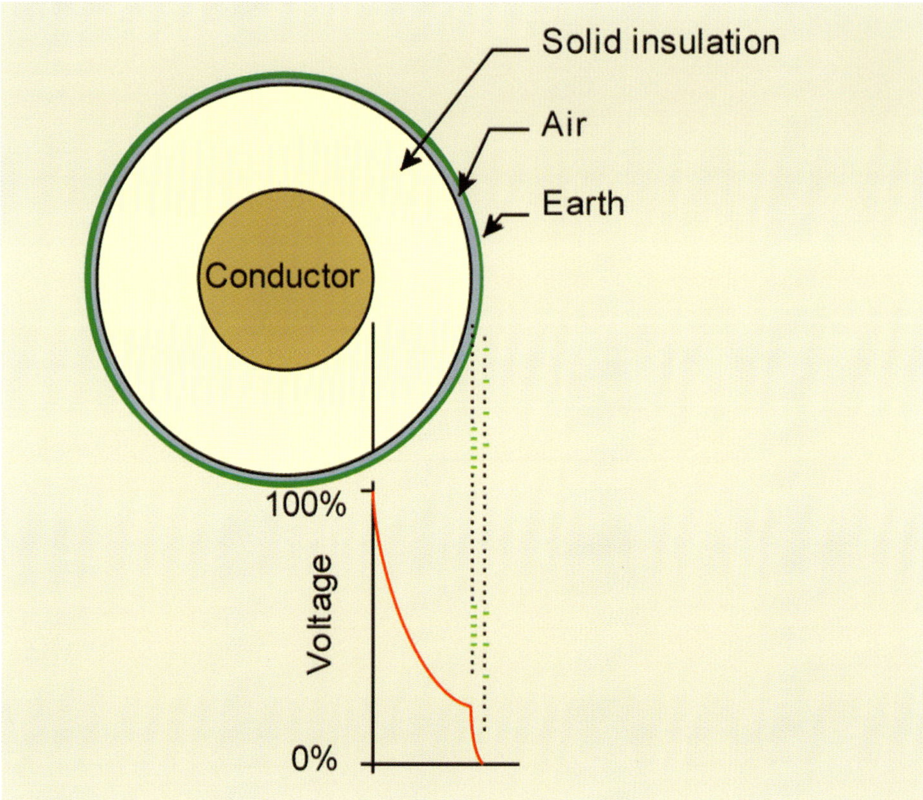

Figure 8
Voltage distribution across cylindrical mixed dielectrics

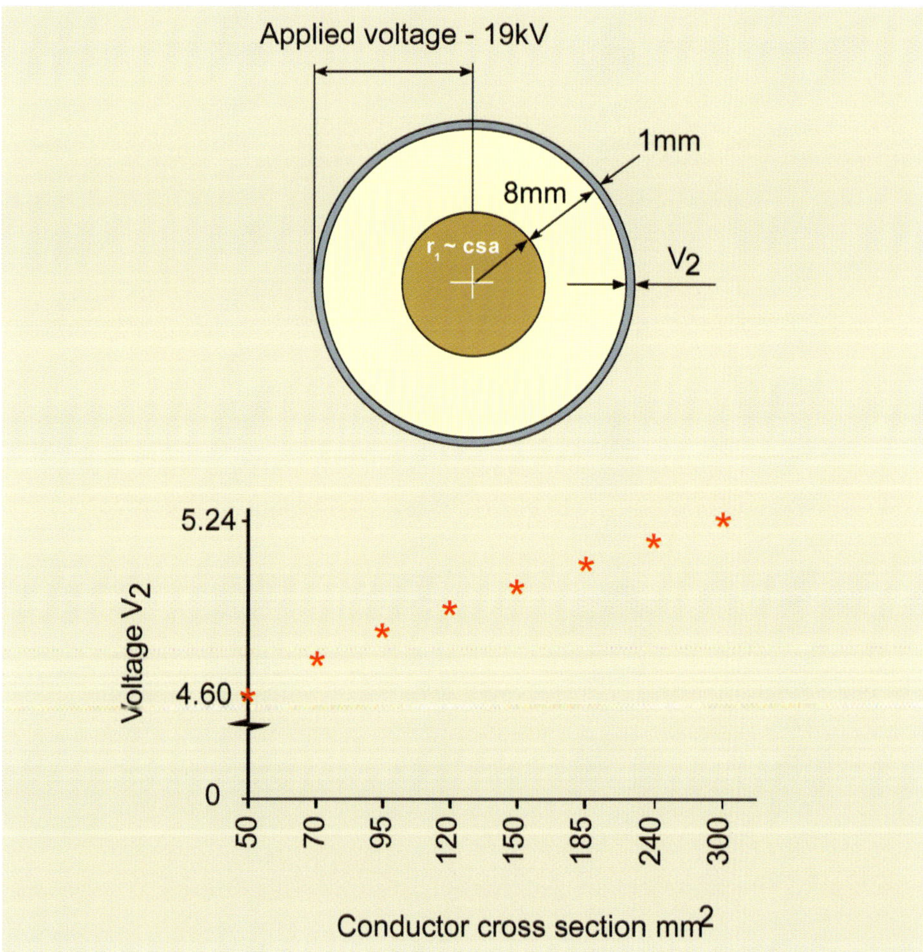

Figure 9
Variation in voltage distribution as a function of conductor csa

The voltage across the air is 4.8 kV rms or 6.8 kV pk, which is over twice the voltage required to break down the air. Unlike in the example of fig. 5, the voltage drop is a function of the inner electrode radius, thus for different conductor sizes, the voltage across the air changes as shown in fig. 9.

It is seen that as the conductor size increases, so the potential stress at the outer surface of the insulation increases. This means that when assessing a cable accessory for partial discharge, the most onerous case is for the largest conductor size. As a result of this type testing is now required at the smallest and largest conductor diameters.

2.3.4. Maximum electric stress in a cylindrical dielectric

It has been shown earlier that the electric stress at any radius, r, is given by

$$E = \frac{V}{r \ln\left[\frac{r_2}{r_1}\right]} \text{ volts/metre}$$

and it can be seen that the maximum stress is when r is a minimum, therefore in the above example of two dielectrics we have the stress at the conductor is

$$E_1 = \frac{V_1}{r_1 \ln\left[\frac{r_2}{r_1}\right]}$$

and at the solid/air interface, on the air side it is

$$E_2 = \frac{V_2}{r_2 \ln\left[\frac{r_3}{r_2}\right]}$$

The value of V_2 has already been calculated for a typical set of values as 4.8 kV rms, therefore $V_1 = V - V_2 = 14.2$ kV rms.

Hence,

$$E_1 = \frac{14.2}{7 \ln\left[\frac{15}{7}\right]} = 2.66\text{MV / m}$$

and

$$E_2 = \frac{4.84}{15 \ln\left[\frac{16}{15}\right]} = 4.96\text{MV / m}$$

We notice two things here, firstly, the stress is much higher in the air than in the solid insulation, and secondly, the stress in the air is much greater that the breakdown strength of air. The peak value of E2 is

$$E_2\text{pk} = 4.96\sqrt{2} = 7.0\text{kV}$$

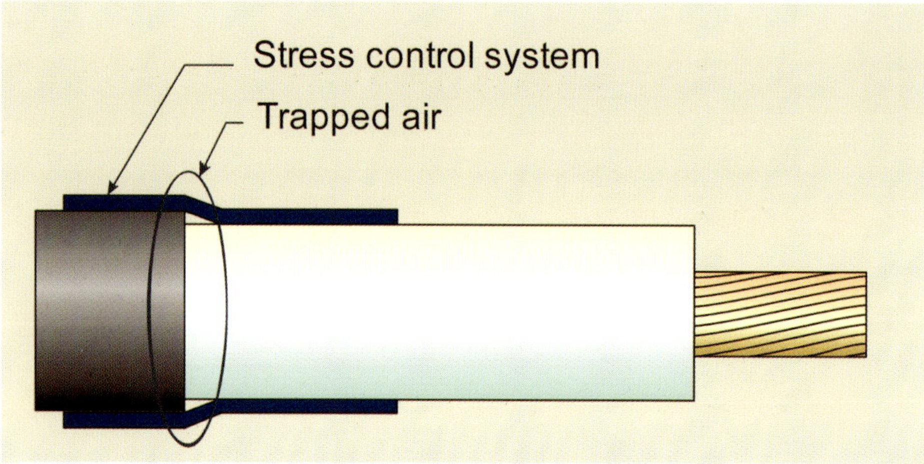

Figure 10
Air/solid dielectric combination at the end of cable screen

thus, since the breakdown strength of air is 3 MV/m the stress at the surface of the solid dielectric is over twice that of air and if no stress control was used, the area at the cable screen cut would discharge at less than half working voltage.

This example is not strictly accurate when applied to a cable, since when the earth screen is removed, there is no defined earth plane just above the surface of the cable insulation and the electric field is distorted as demonstrated in a later chapter. However, reference to fig. 10 will show a situation where the above model is quite close to practice and this theme will be developed later.

The following is an example of a void being created at the end of a conductive tube which was being used as a Faraday Cage over a joint connector. The instructions called for stress relieving tape to be applied to eliminate the void when the insulating tube was applied on top but unfortunately the jointer omitted this.

Figure 11 clearly shows discolouration of the XLPE cable insulation caused by discharges. This phase had not failed but an adjoining one had and as in most failures the clues are to be found in one of the phases which have not failed. The cut on the XLPE was caused during the removal of components to ascertain the reason for failure and was not the fault of the jointer.

This demonstrates the critical issues facing a jointer when building up a joint or termination and has lead to many failures on terminations in particular.

Figure 11
Discharging at the end of conductive tubing

Chapter 3

Some practical examples of mixed dielectrics in accessories

When a cable is prepared for jointing or terminating the earth is removed to expose the cable core insulation. This immediately introduces air into the situation and this has to be addressed to prevent problems. The following defines the areas that must be addressed to ensure satisfactory operation of the accessory. This chapter will simply highlight areas that need to be treated. The methods of doing this will be presented in the following chapters.

3.1. Cable breakout area of paper belted cable

When removing the lead or aluminium sheath and the belt and carbon papers of an unscreened paper cable, there is air surrounding the three cores.

Figure 12 shows a cross section of the cable and the obvious areas that can cause problems.

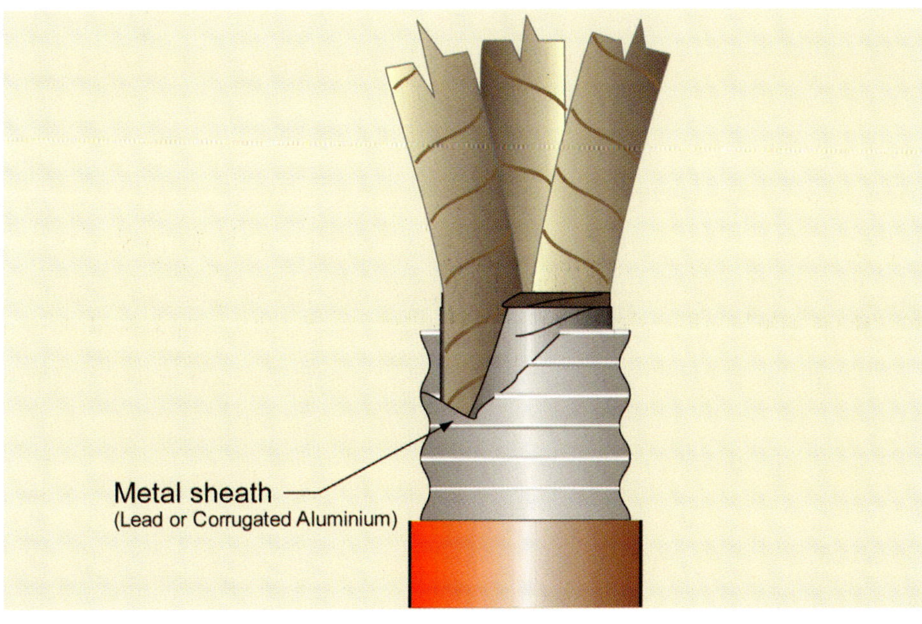

Figure 12
Cross section of a paper belted cable

3.2. Adjacent unscreened cores

When three core cables are to be terminated there is always the possibility that adjacent cores will touch each other in an unscreened area. This is because the roll of the cable is such that when it enters a cable box the core positions are arbitrary in relation to the equipment bushings. In order to achieve correct phasing it is often necessary to cross cores and they will very often touch each other at the point of the cross. This situation is not a problem with joints because the cores will be screened, or as is becoming quite common, resin will be poured around the cores.

The problems caused by core crossing of terminations are worst on paper belted cables, but screened cables can exhibit the same problem if they have been incorrectly prepared (e.g. the screens are removed too close to the breakout point of the cable).

The voltage divider effect in this situation with paper belted cable is shown in fig. 13(a) where a capacitive divider effect is clearly seen between the two cores. If the cores touch, the problem does not go away, as the capacitive effect is merely moved to either side of the contact as shown in fig. 13(b).

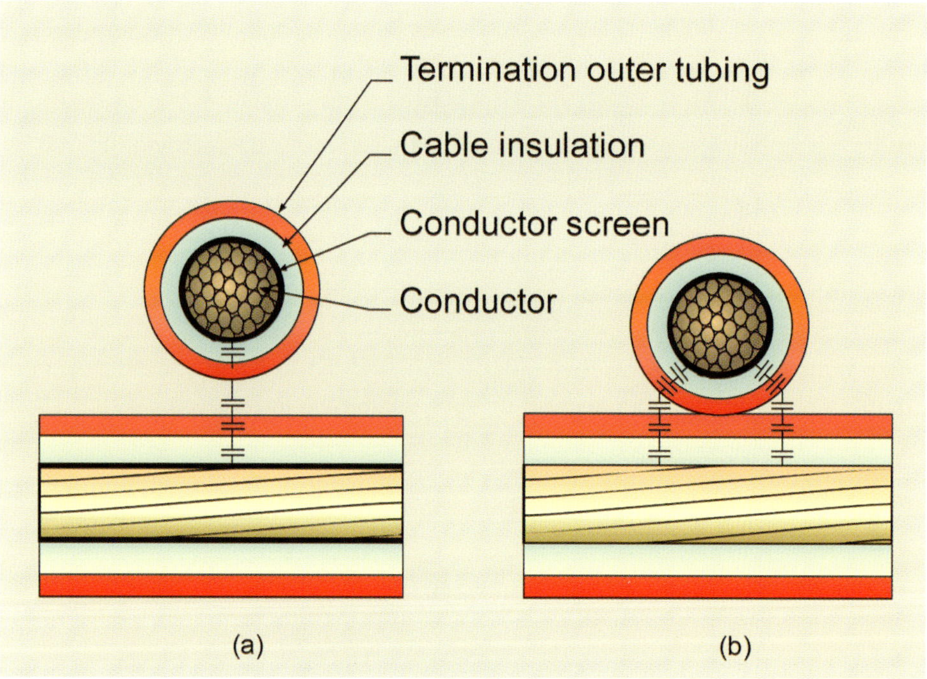

Figure 13
Cross core situation with paper belted cable

3.3. The end of an earth screen on polymeric cable

Whenever a cable is prepared to accept a joint or termination there is potential for air to be trapped as shown in fig. 14. Note that if a knife has cut into the insulation a void is created that is difficult to fill and it also gives rise to a high stress point which may lead to treeing.

At the point of the screen removal there is a capacitive divider effect which leads to a voltage across the air in the region of the end of the screen and the cable insulation. This can be graphically illustrated in a darkened laboratory if a voltage is applied to the cable prepared to accept an accessory. It will be seen that as the

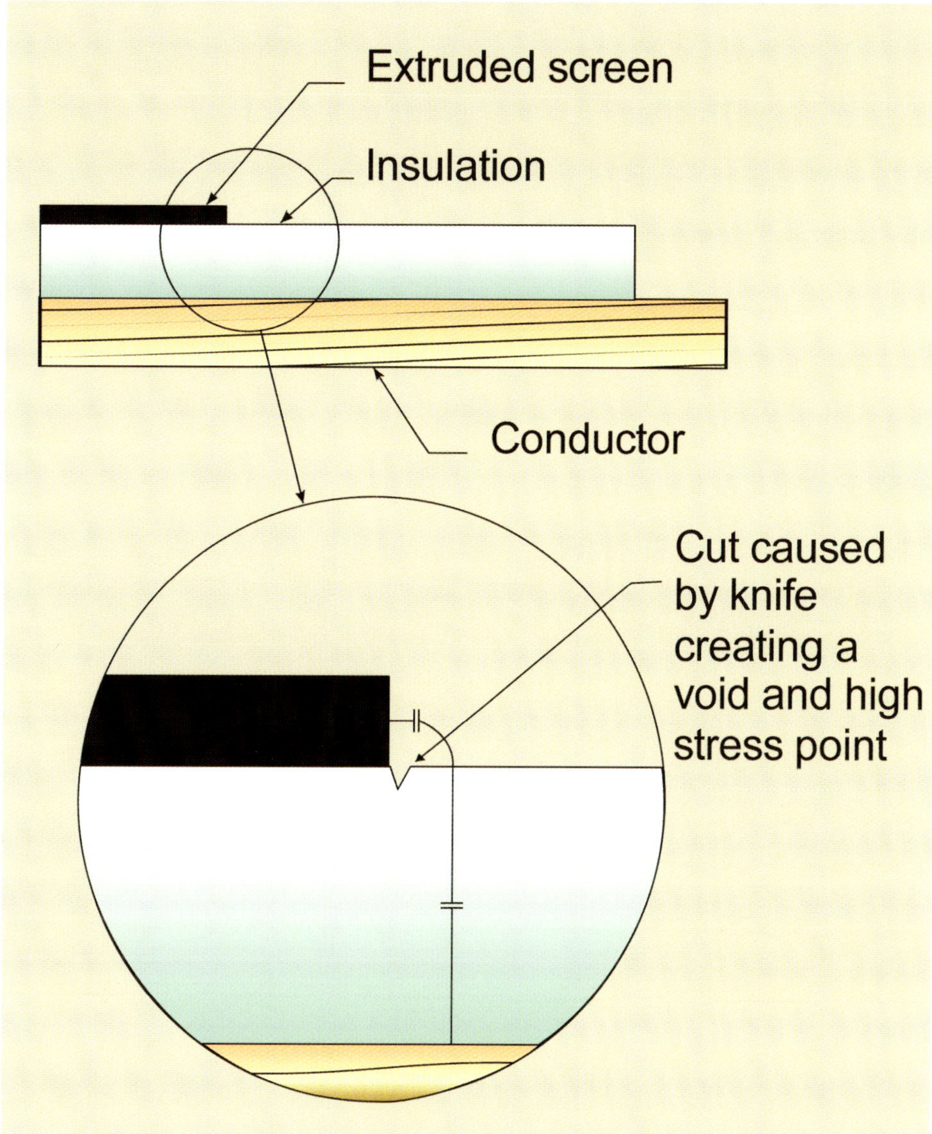

Figure 14
The screen cut on a polymeric cable

voltage is increased a blue glow appears around the screen-cut area at a voltage less than the working voltage of the cable. This glow is the air breaking down electrically and will be clearly visible at about 6 or 7 kV on an 11 kV cable. The same test on a 33 kV cable will exhibit the blue glow at about 12 or 14 kV rms. Partial discharges will be detected at much lower voltages using a discharge detector and will be in the order of 4 kV for 11 kV cable and 8 or 9 kV for a 33 kV cable.

Referring now to the example in the previous chapter where air of 1 mm was positioned around 8 mm of solid as shown in fig. 7 it was shown that just over one quarter of the total voltage was dropped across the air. In other words, an applied voltage of about 8 kV would be sufficient to cause electrical breakdown of the air. This theoretical example is not totally dissimilar to the end of the earth screen shown in fig. 14 above and the result is very similar to the actual voltage at which discharging is detected on a prepared cable. It must be pointed out, however, that the actual calculation for a cable is rather more complicated and 3 dimensional in nature, not the simple 2 dimensional example shown here.

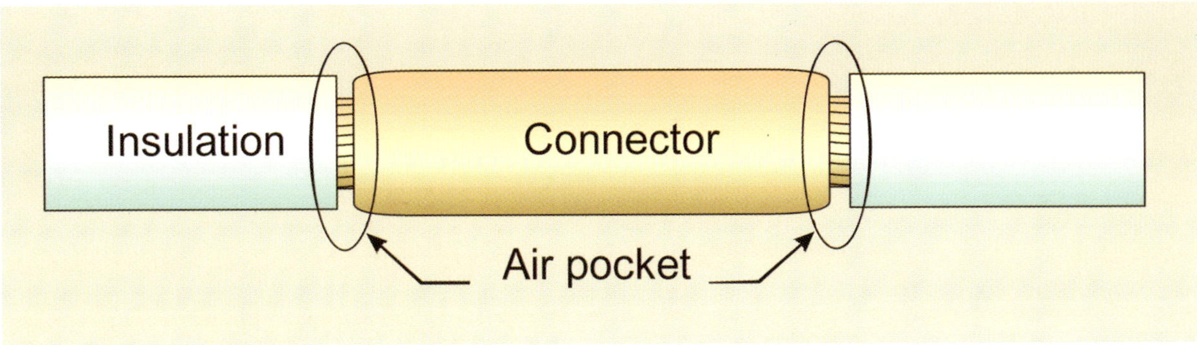

Figure 15
Air pockets at ends of connector

3.4. Area surrounding a connector in a joint

Cable conductors are usually joined either using crimp connectors or mechanical shear head bolt connectors. The type of connector used is immaterial in this present discussion, but will be raised in a later section. Figure 15 shows the area, which clearly is a potential weak spot.

When joints are considered in more detail it will be clear that there are other factors to be taken into account. These include the diameter of the connector in relation to the insulation and any sharp edges that may result from crimping, but this will be covered at that time.

The discussion so far has been of a relatively theoretical nature, but it should be remembered that good cable preparation is essential to the satisfactory operation of accessories. The scenarios referred to above exist no matter how well the cable is prepared, however, poor workmanship can lead to other problems. Therefore before continuing specifically with cable accessories the following two chapters will concentrate on the critical areas of cable preparation especially relating to polymeric cable.

Chapter 4

Some notes on cable preparation

This treatise would not be complete without a few notes on cable preparation. It does not pretend to be a training manual, but may act as an aide memoir to engineers who have some responsibility for cable accessories. This chapter will mainly deal with polymeric cables (XLPE & EPR) concentrating on the most critical area, which is the removal of the two commonest types of earth screen. Before this, a selection of other tools is show. The tools shown are by no means the only ones available but give an indication as to the type of equipment a jointer should have.

4.1. Tools for removing outer sheath and insulation

Figure 16 shows a simple and effective tool for removing the outer sheath of cables. It is fitted with a guarded blade to prevent cutting into wire screens and armours.

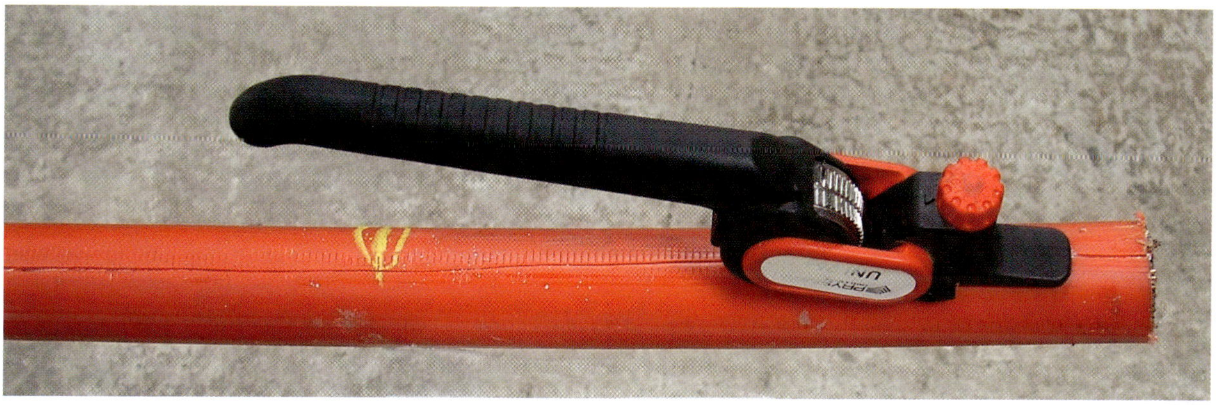

Figure 16
Tool for removing outer cable sheath

More robust tools which perform the same task are shown in fig. 17. In this case there are different sizes for different cable diameters.

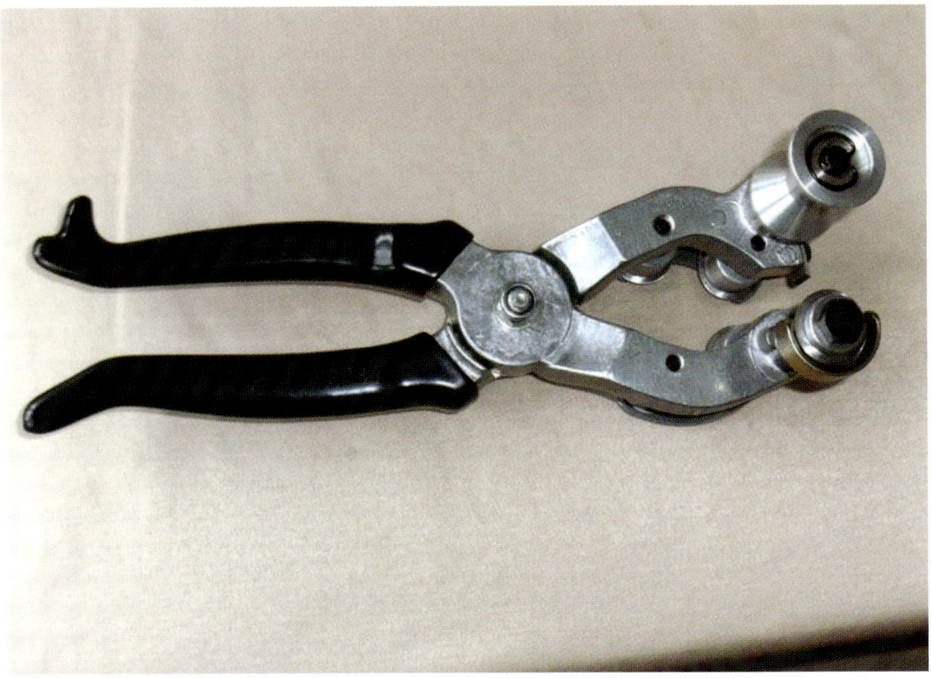

Figure 17
Cable sheath stripping tools

Figure 18 shows two tools for removing the insulation from a cable. The blade is adjusted to a depth which allows the conductor screen and the insulation to be removed without damaging the conductor. The tool on the right has an adjustable stop to allow the correct removal of the insulation to accept a lug or connector.

All the above tools are very useful but the tasks they perform could be carried out using a sharp knife to make the initial cuts and a screwdriver to prise off the jacket and insulation. This is not recommended but could be done in an emergency. However, when the next part of the cable preparation is undertaken, which involves the removal of the earth screen, it is essential to use quality tools.

4.2. Removing a bonded screen

A bonded screen will have to be cut from the cable insulation and there are a variety of tools on the market that will perform this task. Figure 19 shows two typical tools.

Using a correct tool is essential to prepare the cable correctly, but the following points should be noted:

- Ensure the blade is set to the correct depth – excessive removal of the insulation will reduce the amount of insulation and increase the stress in the cable.

- Use silicone grease to lubricate the screen before attempting to remove it.

- Ensure the cable core is straight – a bend in the cable can lead to the screen not being removed around the whole of the cable.

- If small amounts of screen are left on the cable, carefully shave to remove it remove with a piece of glass as shown in fig. 20 or if only small amounts use a fine grade abrasive cloth.

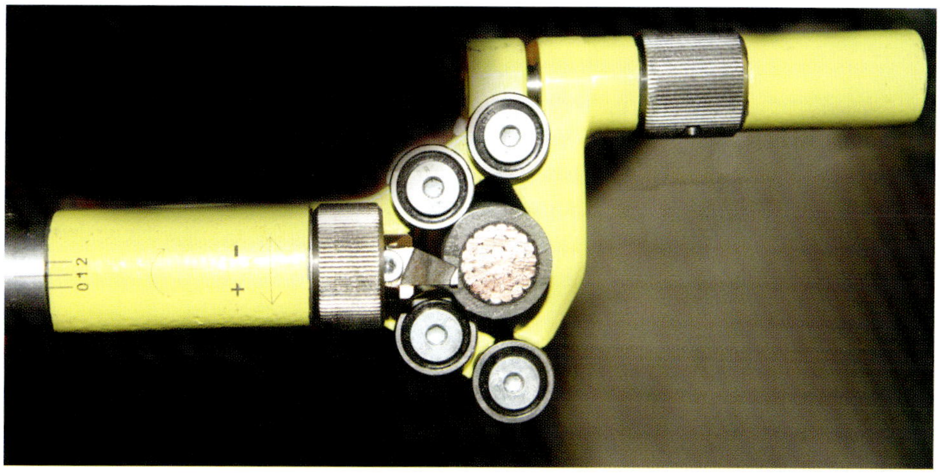

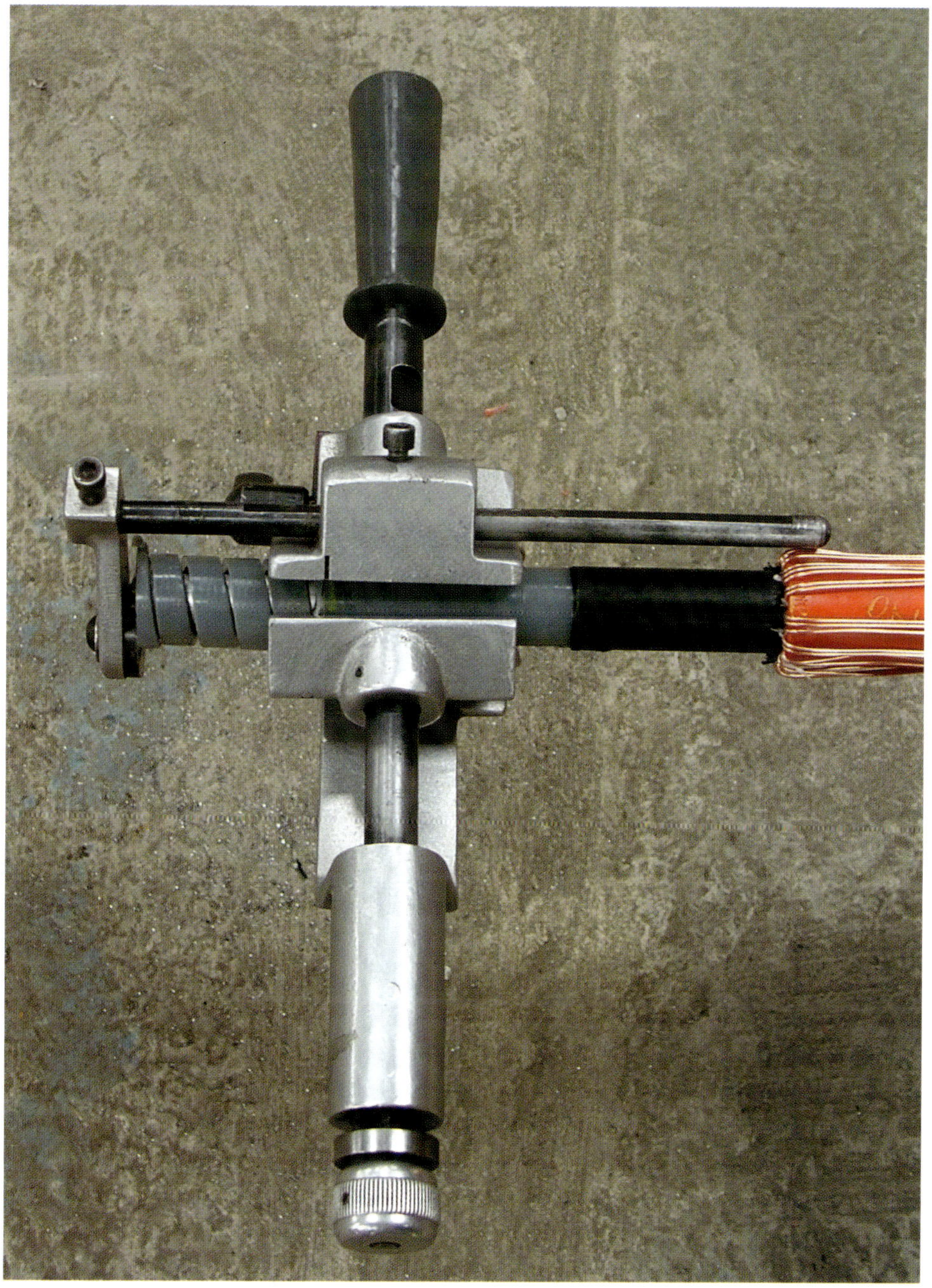

Figure 18
Two tools used
for removal of
cable insulation

Figure 19
Tools for removing bonded screen from cable

The screen which had not been removed totally by the shaving tool demonstrates clearly the need to have straight cable. Often the missed screen is on the underside of the cable and may not be immediately obvious therefore it is essential that close inspection is carried out. Shaving this screen off is quite easy using a piece of glass as shown.

4.3. Removing easy peel screen

Many failures that occur in practice are caused by careless removal of the screen and in particular the method of removing the screen at the cut-off. Many jointers

Figure 20
Removal of screen missed by shaving tool due to bent cable

Figure 21
Filing through the screen at cut-off point

still use knives and cut through the screen and into the insulation. This effectively creates a void similar in shape to the point-plane configuration shown in fig. 2. The result is that partial discharges start and gradually break down the insulation of the cable. Unfortunately, this is not indicated during pressure testing prior to commissioning and can take months or years to cause breakdown, depending upon the depth of the cut.

There are various tools available to help in the removal of this type of screen.

4.3.1. Method – using basic tools

The tools required here are inexpensive and can prove perfectly satisfactory with practice. Furthermore three core cables can be worked on readily, which is not always the case with more sophisticated tooling.

The most critical part of this process is ensuring a clean break at the end of the screen. This is best done by using a round file at the cut-off point, having first marked the point with binding wire or constant force spring which will then act as a guide for the file – see fig. 21.

Using a file to just cut through the screen to expose the insulation is the safest method to use. As with all steps of cable preparation, the job should not be hurried, but be carried out with care. No matter how good the termination or joint, if the cable insulation is cut during preparation, failure is most probably going to occur.

Figure 22
Scoring the screen longitudinally using tool with pre-set blade depth

Figure 23
Peeling a strip of screen from the cable core

Having just exposed the screen around the circumference of the cable, the next step is to cut into the screen along the length of the cable as shown in fig. 22. Five or six such cuts, equally spaced, should be made (this is often referred to as *tram lining*). A tool with pre-set blade depth should be used – these are set to 0.4 mm and 0.6 mm. Always check the cable screen thickness to ensure that the blade does not penetrate the insulation. Generally the 0.4 mm depth is for 12 kV cable and 0.6 mm is for 36 kV cable. The use of a knife will probably lead to the insulation being cut, again with dire consequences, and is not recommended.

Having made the cuts in the screen, strips of the screen should then be peeled from the end of the cable to the file cut. Removing the screen initially may be a little awkward, but the use of a small screwdriver or knife will prise it off at the end where the lug or connector is to be installed and then it can be peeled as shown in fig. 23. Repeat with each strip until all have been removed and then remove the binding wire marker.

4.3.2. Method 2 – using sophisticated tooling

The use of a proprietary tool as shown in fig. 24 is very effective on single core cable but is not always suitable for three core cables if the screen has to be removed close

Figure 24
Tool for scoring easy peel cable

to the breakout of the cores because there is insufficient separation between cores to allow the tool to rotate. This particular tool has a blade depth adjustor and the screen can be scored longitudinally, as shown or it can be spiralled. The latter method allows the screen to be removed in one piece rather than in several with the tram line method. If the cable conductor is sector shaped obviously the tram line method is the only viable one.

4.4. Graphite coated conductive tape wrapped cores

This cable construction was not unpopular several years ago but now has been superseded by the two types discussed above. It is mentioned here simply because there will be some still in service and it may be necessary to re-terminate it or join to new cable at some future date.

The conductive tape should be treated in the same manner as a copper tape, but extra care is required to prevent it unwinding and should therefore be held in place with adhesive tape or carefully applied thin binding wire. The graphite layer should be removed using a solvent and care must be taken to leave a smooth edge about 15 mm beyond the tape. It is usually necessary to use several clean cloths to gradually clean the cores and remove all trace of graphite.

Treating the end of any screen cut is the next vitally important step and will be dealt with in the following chapter.

Chapter 5

Treating the screen cut on a polymeric cable

One of the most common causes of failure of polymeric cable accessories is breakdown of the cable insulation at the end of the screen cut. This is the result of internal partial discharges brought about by one or more of the following:

- *poor cable preparation*
- *poor installation*

which can result in:

- *cuts in the primary insulation*
- *the presence of voids*

which in turn causes:

- *partial discharges*
- *ultimate failure*

5.1. The problem area

Correctly prepared cable still presents an area of potential problem at the screen cut – see fig. 25, especially with easy peel screens. This area is a site for voids and

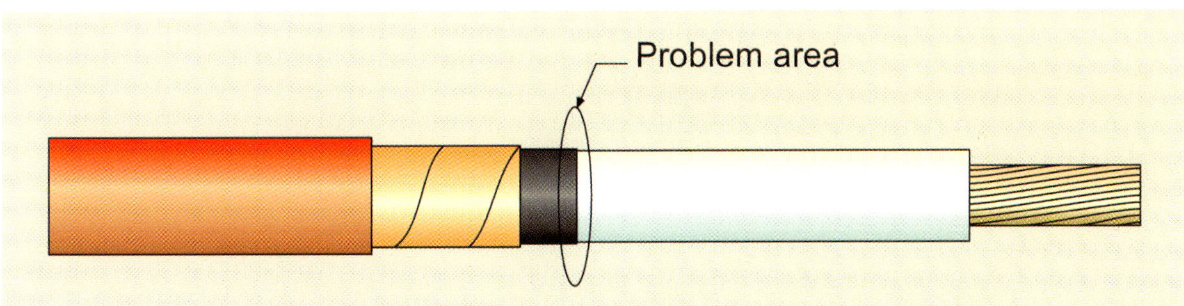

Figure 25
The end of the screen on a polymeric cable

must be treated correctly because the stress control system usually does not exactly follow the profile at this point. Several techniques have been devised over the years and the most common ones will now be discussed.

Note that the copper tape screen is removed about 10 mm from the end of the extruded screen layer. If the cable is wire screened the wires are bent back at least 25 mm from the end of the screen cut.

The use of the correct tool to remove the screen does usually present a chamfered edge, but this may not be sufficient to prevent void formation. In any event, it is preferable to use one of the following techniques as they will help address potential problems caused by poorly removed screens.

5.2. Some techniques for eliminating potential voids

5.2.1. Conductive paint

The use of conductive paint as shown in fig. 26 extends the screen layer and eliminates the step at the original screen cut. This technique will not lead to void formation at the end of the earth plane because the thickness of the paint is minimal. This technique is perfectly satisfactory, but the paint can dry out during storage making it impossible to apply so there is a degree of skill required in applying it.

It is usual to smear a thin layer of silicone grease on the cable insulation and onto the paint prior to fitting the stress control system. This helps fill any minor surface irregularities caused by less than perfect cable preparation.

The use of conductive paint is now rarely used at system voltages up to 36 kV, but at higher voltages is often used in conjunction with other systems.

5.2.2. Void filling stress relieving tape

Tape, which is carefully wrapped around the area of the screen cut will provide a void free smooth profile – see fig. 27. This method remains the most popular at the present time.

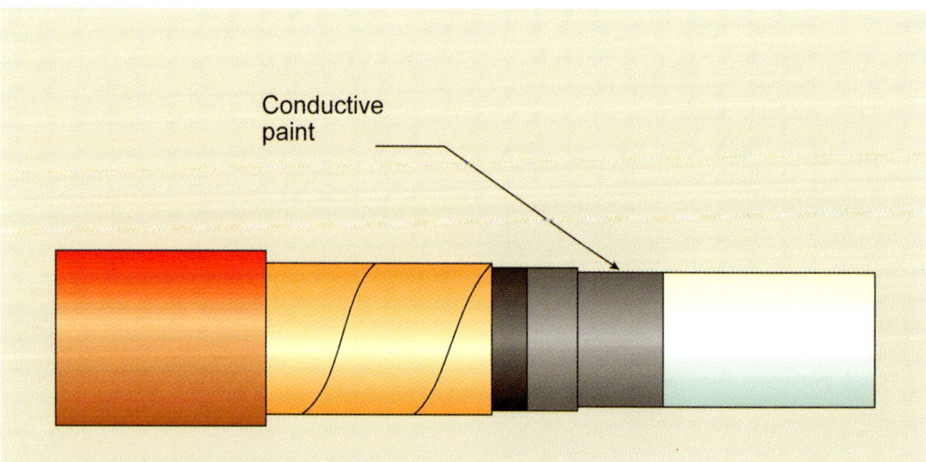

Figure 26
Conductive paint applied to the end of the screen cut

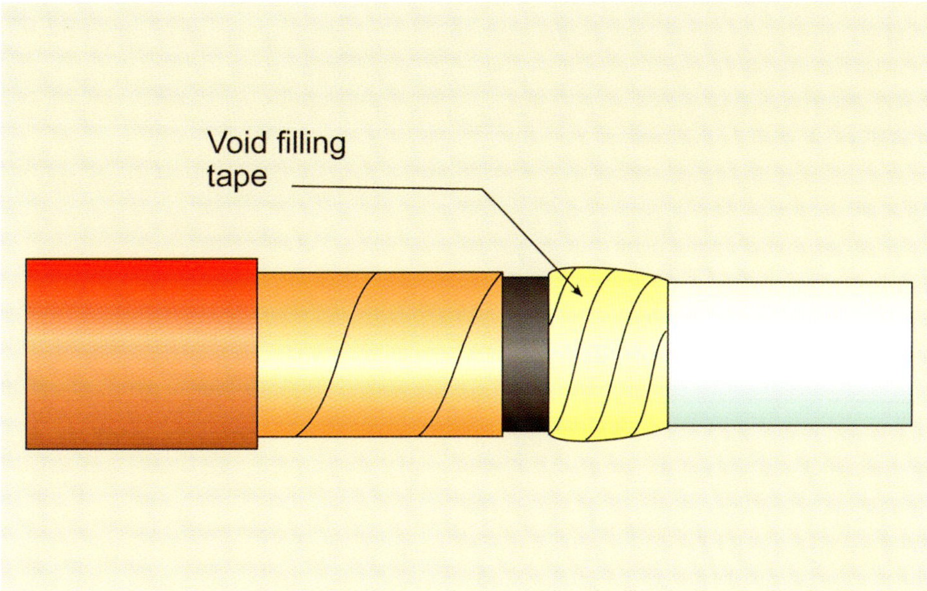

Figure 27
Void filling tape wrapped around the screen cut

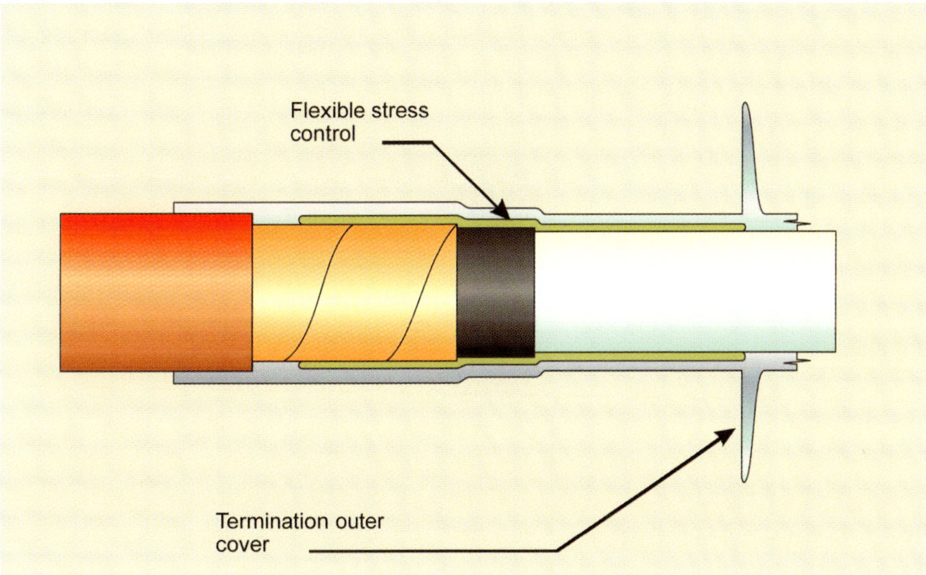

Figure 28
Flexible stress control incorporated in the accessory

Several modern accessories contain stress control, which is an integral part of the system. This can be in the form of an elastomer which is sufficiently pliable at the installation temperature that it will follow the profile of the screen cut. A coating may be applied within heat shrinkable tubing which softens during installation. Another type is an elastomeric patch, which is wrapped around the cable core and over the screen cut. Figure 28 shows how the product fills in any voids and irregularities at the screen cut and eliminates the need for other treatments such as conductive paint or void filling tape.

5.3. Pros and Cons

It is fair to say that all the techniques described above will work perfectly satisfactorily if applied correctly but there is an element of skill required and care should be taken at this most important stage.

The first two methods do not give stress control in their own right. Although the void filling tape may be stress relieving, a stress control system must be installed subsequently. The most common causes of problems are poor application and the jointer forgetting to apply the paint or tape.

In theory the flexible stress control system incorporated in the body of the accessory has a distinct advantage in that it does not require treatment of the screen cut. However, the application must be carried out in the correct manner, otherwise air may be trapped and resulting voids can give problems, especially at higher voltages.

Chapter 6

Stress control on paper belted cables

It is accepted that the use of paper belted cable has been virtually eliminated but there are huge amounts in the system and if it has to be re-terminated or jointed to polymeric cable the following sections will be relevant.

6.1. Initial preparation of paper belted cable

When the outer metal sheath has been removed black carbon papers, which are wrapped around the three cores, are exposed. These papers are removed to within 10-15 mm from the metal sheath – this dimension will be specified by the accessory manufacturer. This reveals the belt papers which are the same colour as the core

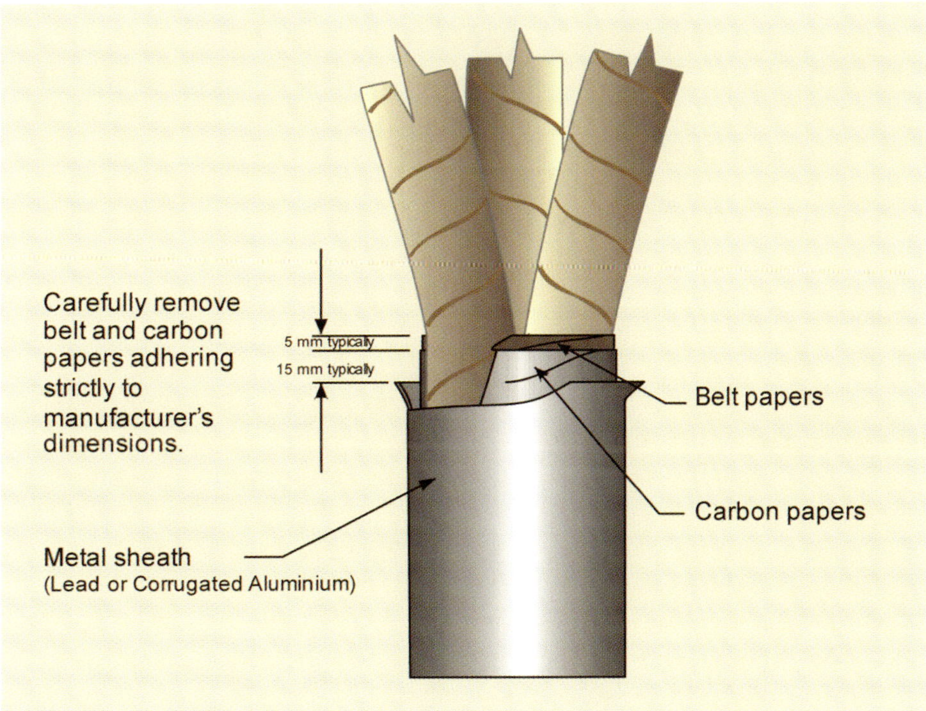

Figure 29
Paper belted cable in area of metal sheath removal

insulation. The belt papers are then removed a further 5-10 mm beyond the carbon papers as shown in fig. 29.

Many failures occurred in the early stages of developing dry teminations due to air being trapped in the area shown in fig. 29. Most of the time the failure was from phase to earth but another problem arose due to the first layer of heat shrink tubing not shrinking completely, which in turn created voids.

The problem arose because the tubes were pushed as far down the cable as possible but then the flame used to carry out the shrinking did not penetrate between

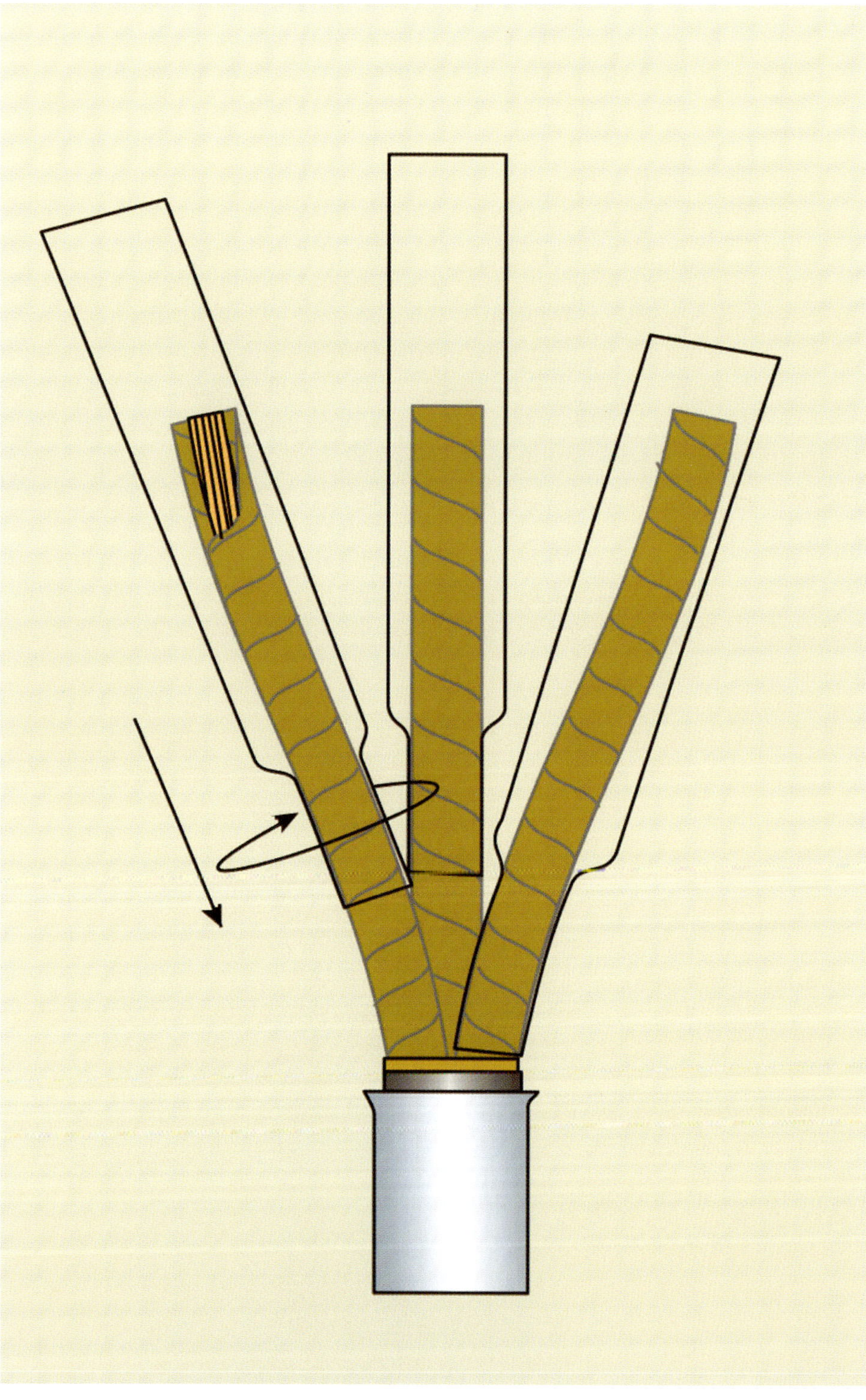

Figure 30
Correct application of inner tubing on heat shrink termination

the cable cores and the tubing failed to fully shrink. To avoid this problem it is necessary to initially shrink the tubes as shown in fig. 30, wait until they are cool then twist in the direction of the cable papers and slide down into position.

6.2. Cable breakout area of unscreened paper cable

It was mentioned in an earlier section that the cable breakout area of an unscreened paper cable contains areas where air could be trapped when using a dry joint or termination. Reference to fig. 12 in chapter 3 shows the potentially problematic sites. Before the advent of dry terminations and joints this type of cable would be immersed in bitumen compound, which pushed all the air from this region.

The solution to this problem with modern cable accessories is to use a plug to fill the areas between the three cores at the break out and a self amalgamating tape wrapped around the three cores and onto the metal sheath. The tape will build up a profile onto which the cable breakout or other device can be installed without trapping air as shown in fig. 31.

The tape used may be stress relieving, but a good quality self amalgamating insulating tape such as butyl is equally as good. The important thing is to eliminate air from the critical regions of high stress.

Once the stress control tubes or preferably semi conducting tubes (see later) are fitted, the breakout is pulled down onto the cable and fits tightly to the void filling tape as shown in fig. 32.

6.3. Screening the cable cores

The need to screen part of the belted cable cores is essential when the cores of a termination have to be crossed. Many failures have resulted due to core crossing of

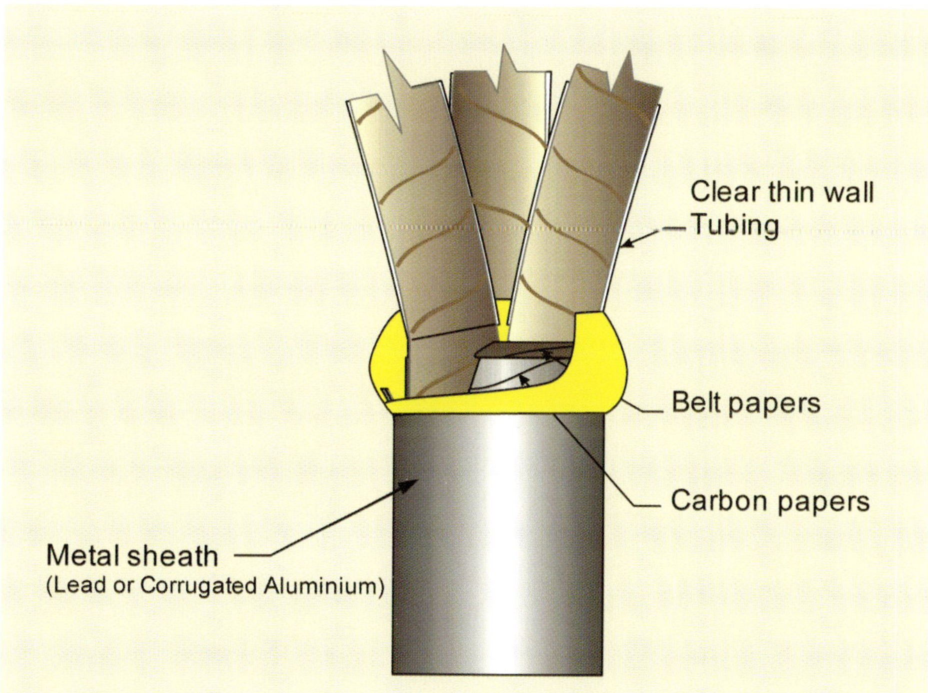

Figure 31
Treatment of paper cable prior to installing breakout

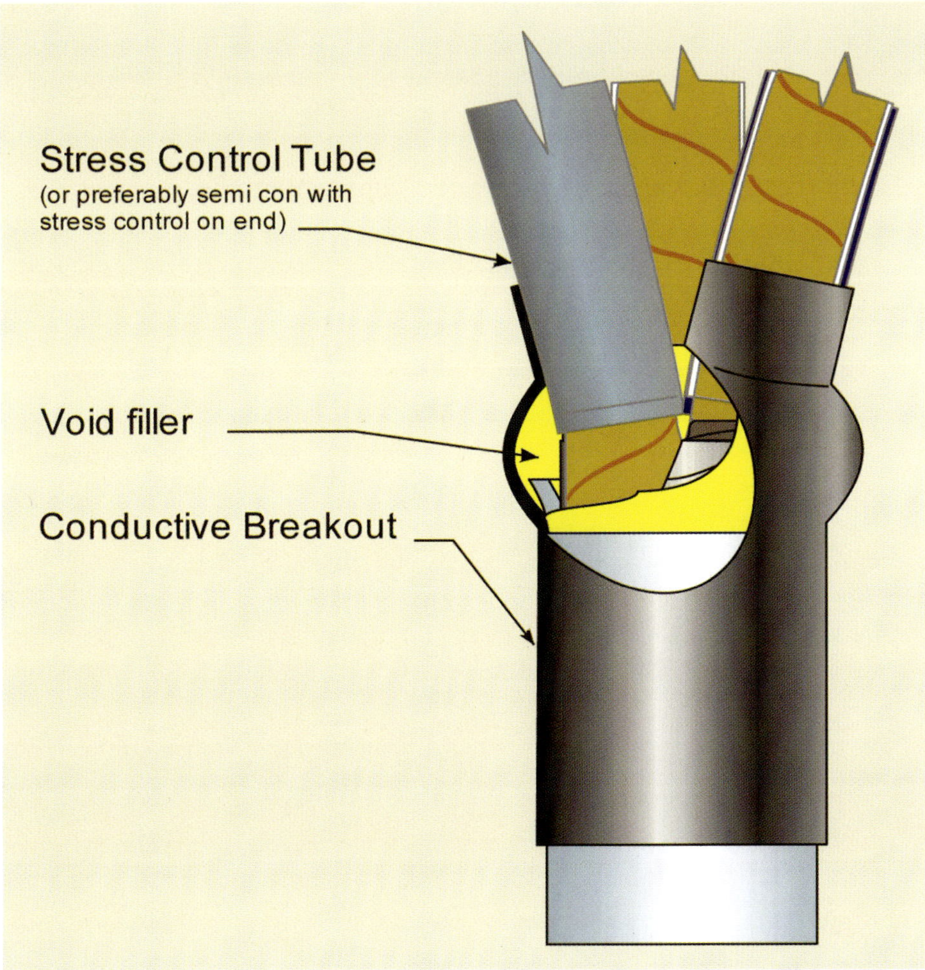

Figure 32
The breakout area with conductive breakout fitted

paper belted terminations as discharging occurs in the stress control area between cores. The same is necessary when installing a joint.

Paper belted cables can be converted into screened cables from the point of sheath removal quite easily but is not an option offered by all termination manufacturers. This modification eliminates the problem often found particularly in terminations where core crossing occurs and unscreened areas of the termination are close or touching. The practicalities of terminations in cable boxes will be considered in more detail in a later chapter. Figure 33 shows a typical arrangement for screening paper belted cable which is standard practice in a joint, but not on terminations.

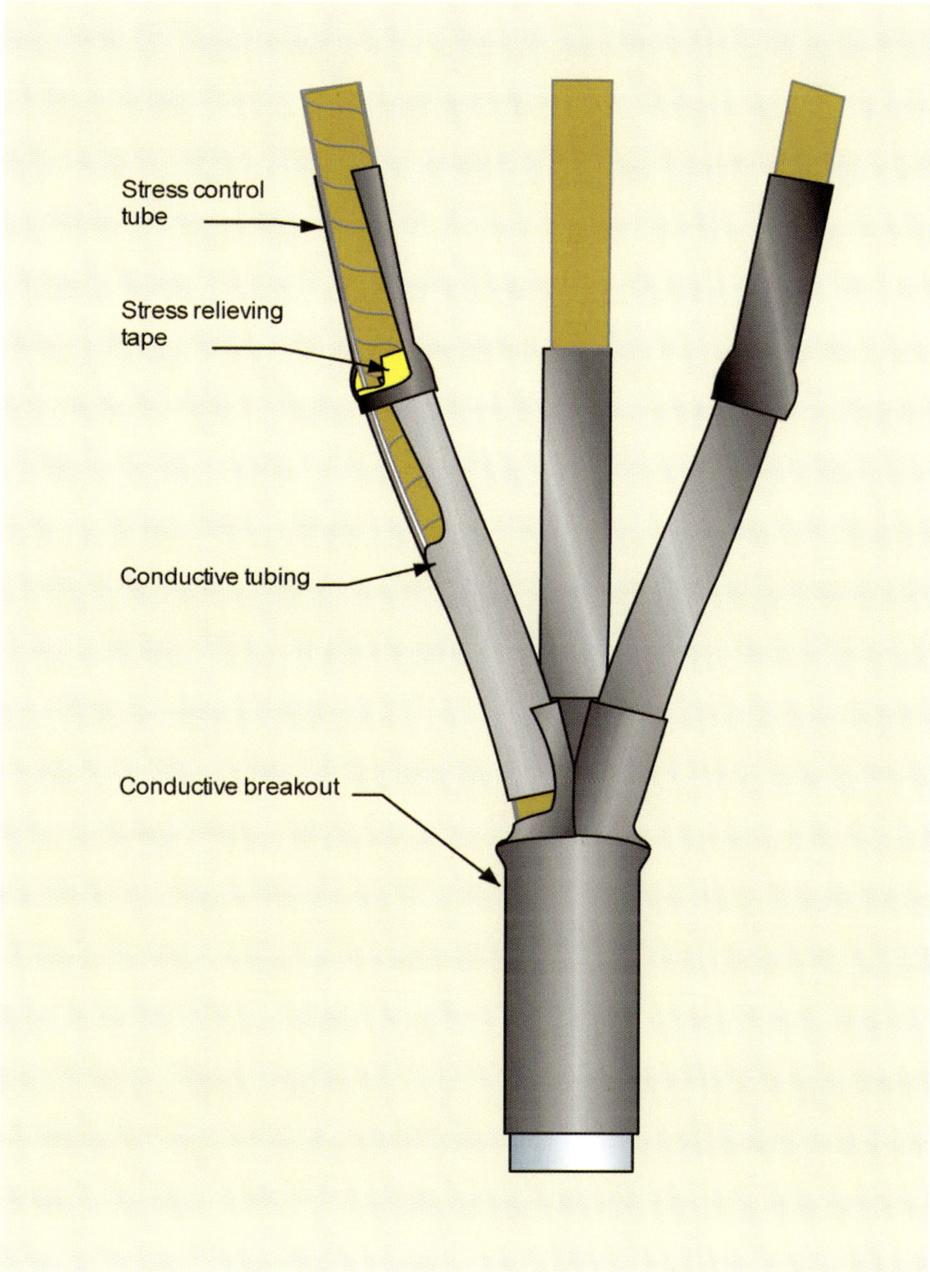

Figure 33
Screening paper
belted cable

Chapter 7

Stress control in polymeric cable accessories

Preparing a cable to accept an accessory will introduce high electrical stresses which have to be controlled. This chapter looks at the causes of the stress and necessary control. The previous chapter examined the construction of a paper belted cable termination or joint up to the point of screening the cores. The high stresses around the area of the metal sheath removal are rendered ineffective by replacing the air with high dielectric strength tape. Remember the stresses are still present and high enough to cause air breakdown but not high enough to have any effect on the materials used to eliminate the possibility of air.

The following will be relevant for screened cables of all types but the illustrations will show polymeric cables only.

7.1. Stresses within a cable

Within the screened area of a cable the stresses are distributed regularly and are such that the cable insulation can easily cope as shown in fig. 34. The annular rings

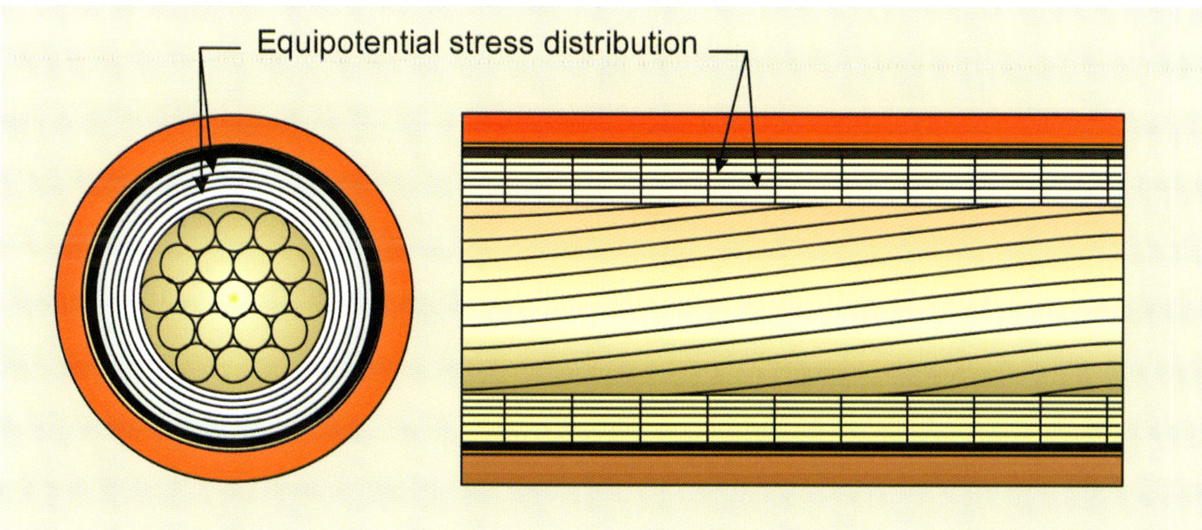

Figure 34
Stress distribution in a screened cable core

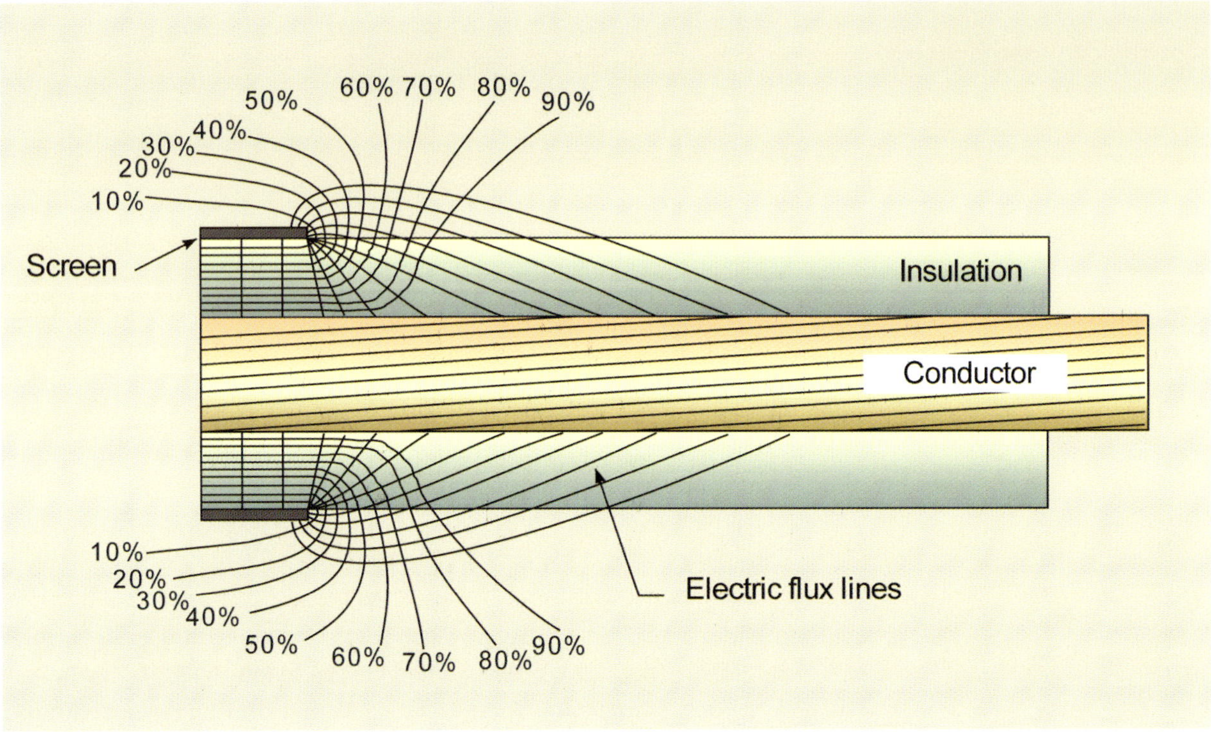

Figure 35
Prepared cable prior to fitting stress control

on the end view are distributed in a manner explained in chapter 2. The side section on the right shows the longitudinal stress lines, which are concentric and thus parallel to each other.

When the earth screen is removed from the cable this uniform stress distribution is totally altered as represented in fig. 35. It is seen that the stress lines are very close together near the screen cut and this leads to high, localised stress.

Figure 35 shows the electric stress lines at the screen cut (or the end of a screen formed on paper belted cable) prior to any stress control being applied. Note that the flux in this region dictates the stress distribution. The equipotential lines are at right angles to the flux.

The effect of removing the earth screen is two fold:

- The voltage appears not only across the cable insulation but also the surrounding air
- The equipotential lines are no longer distributed uniformly and where these lines are close together the electric stress will exceed the breakdown strength of air near the point of earth removal
- On 11 kV polymeric cables at rated voltage (6.35 kV to earth) electrical discharging around the end of the earth screen begins at between 3 & 4 kV
- On 33 kV (19 kV to earth) polymeric cables, discharge inception occurs at about 7 to 8 kV

If the cable was left in this condition the following would occur:

- The discharges will attack the cable insulation
- Ozone will be produced
- There will be a total breakdown of the cable insulation from the earth screen cut to the conductor due to a mixture of the surface attack by the discharges and subsequent treeing

If the cable is situated in an enclosure (e.g. cable box) the ozone will lead to severe degradation of the metal work and surrounding insulation.

7.2. Stress control at the screen cut

There are three ways of dealing with high stress, namely

- Swamp the high stress region with a material having a higher electric strength than air and permittivity closer in value to the cable insulation than air
- Make a geometric shape to encourage a more uniform field
- Use a high permittivity material with a resistivity somewhere between insulating and conducting

7.2.1. Swamping

This technique was used on paper belted cables before the advent of dry cable accessories using bitumen or oil and is used today in the breakout region of paper belted dry systems and some hybrid joints. The theory behind this technique is that in the vicinity of a highly stressed region, a solid or liquid with a high enough electric strength so as not to be unduly stressed, replaces any air that could be present. At 11 kV no other action is required but at higher voltages other solutions are necessary. It will be noted that this technique is employed on dry terminations and joints at the cable breakout of three core paper belted dry accessories – see fig. 32 in chapter 6. Some hybrid joints use resin to perform the same task.

At voltages above 11 kV the stresses at the end of the screen increase considerably. This can cause breakdown of the cable insulation or surrounding solid insulation. Therefore a more sophisticated approach is needed as explained in the next two sections.

7.2.2. Geometric stress control

In order to reduce the stress it is necessary to force the stress lines further apart. This is achieved using a geometric shape in the form of a moulded earth cone within an

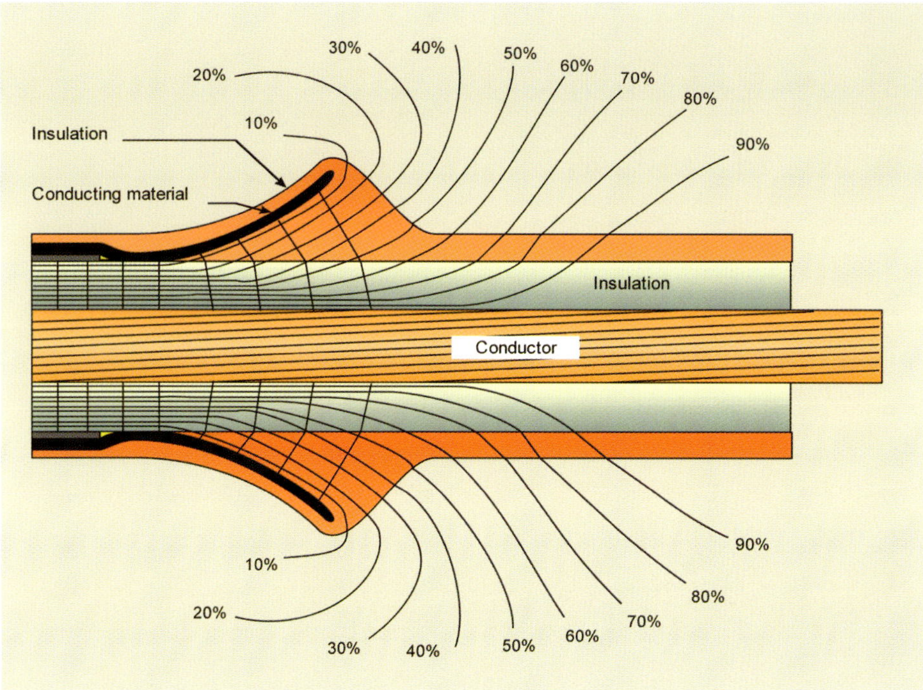

Figure 36
Geometric stress control

insulated jacket as shown in fig. 36. Screened separable connectors, which are discussed in a later chapter, use a similar method to control the stress distribution.

The moulded form of the earth layer in the stress cone is such that it physically encourages the stress lines to space themselves more uniformly and it is a very effective system.

7.2.3. **Electrical stress control**

The geometric solution is fine for single core polymeric cable, but a more versatile system is a stress control tube or patch. This has the advantage of achieving a lower profile.

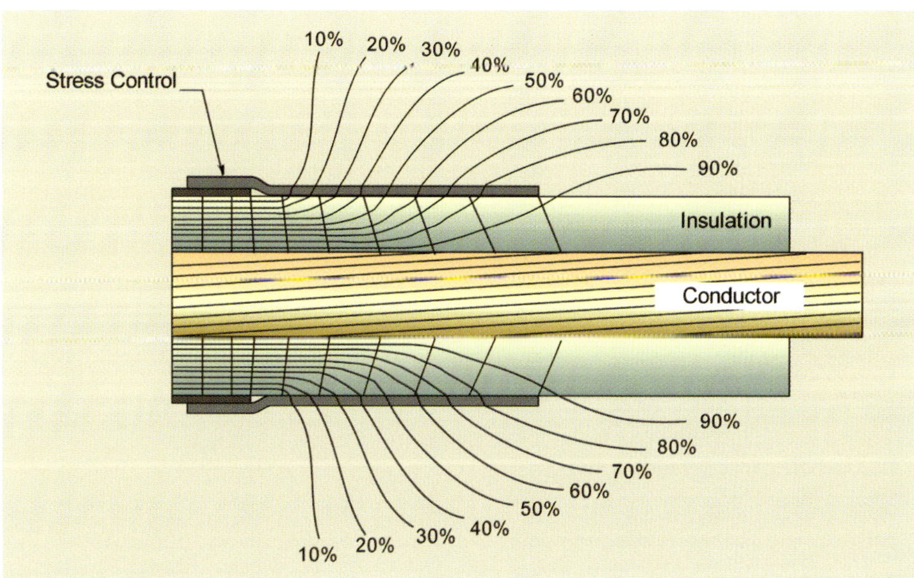

Figure 37
Prepared cable with stress control tube fitted

The stress distribution for this system is shown in fig. 37. It is achieved using materials which have a high permittivity and relatively low resistivity to encourage the capacitive currents to flow along the stress control material. It will be seen that with the addition of the stress control, the current flow and flux through the insulation is changed. This leads to a more uniform distribution of the stress lines and thus greatly reduces in the electrical stress in the area.

With the addition of good stress control the voltage required for discharge inception on an 11 kV cable will increase from about 3 or 4 kV to about 15 kV.

Two points need to be reiterated when discussing stress control. Firstly, it is important that no void occurs at the screen cut and this is achieved using a method explained in a previous chapter. The situation is improved if the end of the screen is chamfered since it is easier to eliminate a void at a chamfer than a square edge. If the cable screen is easy peel, using a round file to create the screen cut will result in a chamfered end to the screen. If the cable screen is bonded, then the use of the correct tool will also ensure a chamfered end to the screen.

Secondly, if a knife is used to mark the screen cut it is essential that the cut does not penetrate the primary insulation. If this occurs then treeing type breakdown as shown in fig. 3 can result. This cause of failure is the commonest found in practice.

7.3. Stress control in joints

In additional to the stress at the screen cut there is a further high stress region in joints and that is over the connector and there are two basic techniques to control this stress as follows:

7.3.1. Stress relieving tape around the connector

In this case stress relieving tape is applied over the connector as shown in fig. 38 to eliminate any air which may subsequently be trapped as the joint is built up.

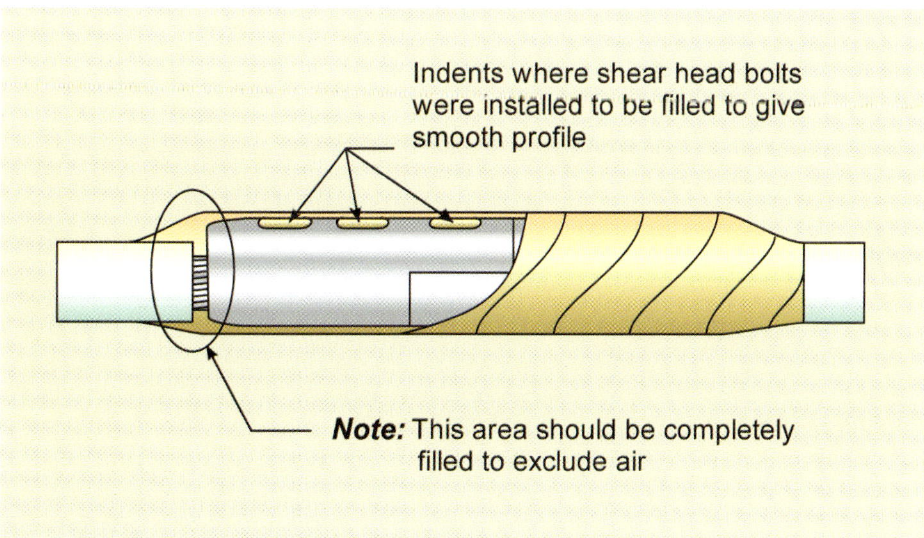

Figure 38
Taping over a connector

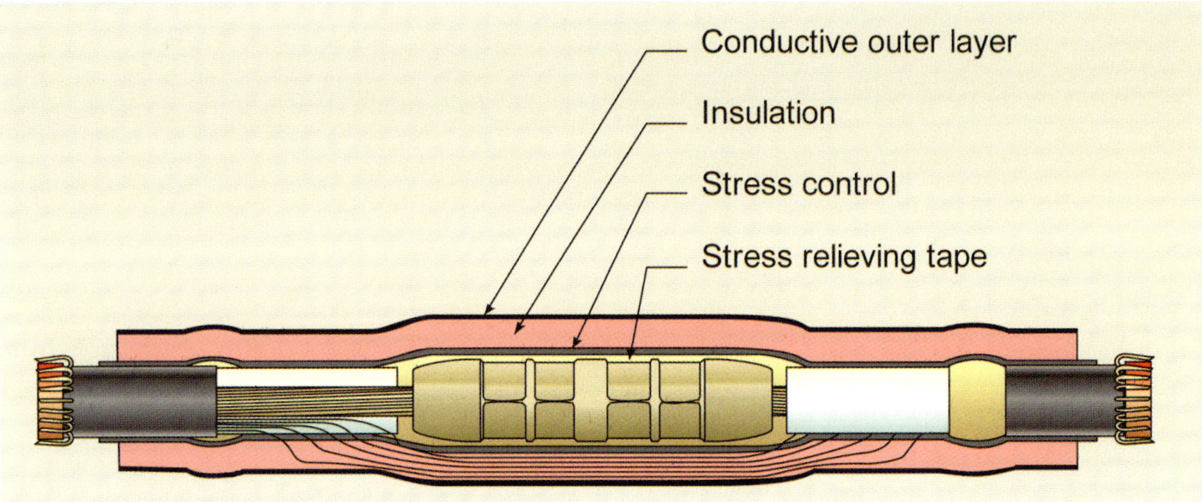

Figure 39
Cross section of a joint

The connector shown is thicker on one side to accept the shear head bolts but there are connectors available now with a symmetrical radial profile which reduces the amount of taping required. The material can also be supplied in the form of a single sheet.

Once the tape is applied a stress control tube is applied across the joint as shown in fig. 39. Note the different type of connector, which is the alternative to the mechanical one shown in fig. 38.

The connector shown in fig. 39 is the hexagonal crimped type and care must be taken when using this product to ensure there are no sharp edges left by the crimping process. Clearly the mechanical connector has an advantage in this regard although both types can be used successfully.

The stress equipotential lines are shown on the lower half of fig. 39 indicating how the stress is equally distributed across the joint.

7.3.2. The Faraday cage

This system eliminates the need for taping over the conductor as shown in fig. 40. The Faraday cage consists of a conductive material which is moulded into the body of the joint and provided it is in contact with the connector at some point, it will

Figure 40
Faraday cage over a connector

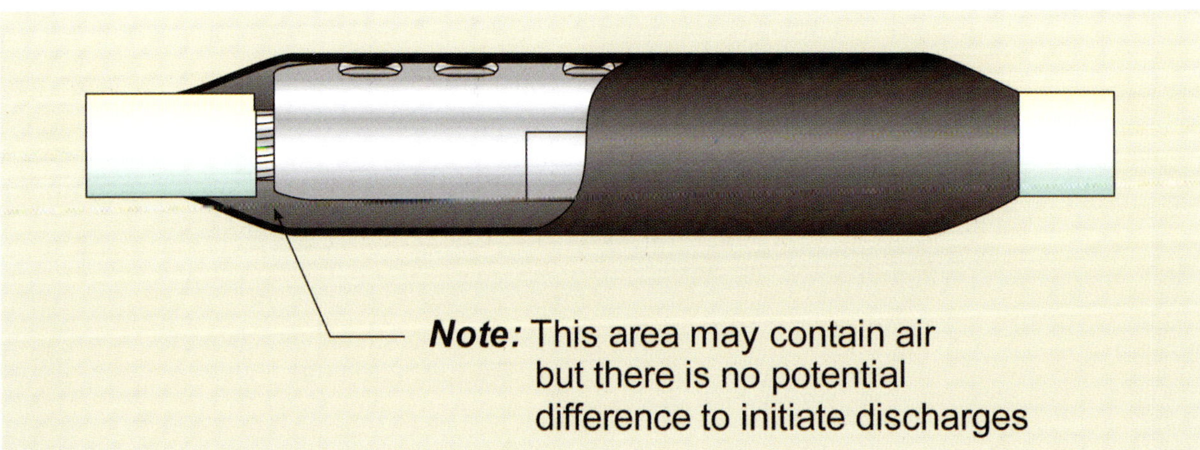

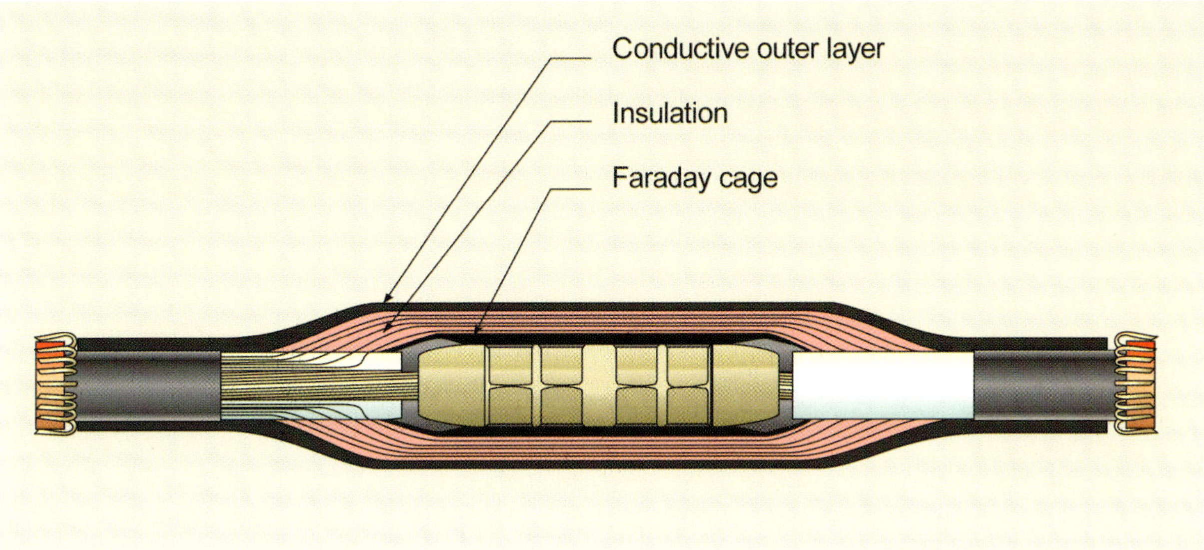

Figure 41
Single core joint
with Faraday cage

assume the potential of the connector. Any air between the connector, conductor and the cage will not be under any stress and therefore will not create a problem.

Joints incorporating a Faraday cage will generally be utilised in cold applied or push on joints. The cross section of such a joint showing the stress equipotential lines is shown in fig. 41.

The Faraday cage is also utilised in screened separable connectors and this will be discussed in detail in chapter 9.

This concludes the section on stress control and the following three chapters will examine termination construction in more detail.

Chapter 8

Terminations 1
A practical appraisal

8.1. The function of a termination

Terminations serve a variety of functions as listed below. Note that screened separable connectors also perform the functions listed below and will be discussed separately in the next chapter.

Function	How achieved
1. To exclude air in critical areas	Use tapes or putties to ensure adequate spacing between unscreened areas.
2. To control electrical stress	Use of geometric moulding or high permittivity material.
3. To provide protection of cable insulation from the environment	Use tubular and moulded shapes of either push-on, heat shrink or cold shrink materials with good anti-track and erosion properties, and, for outdoor terminations, UV stability.
4. To provide an earth connection	Most earth connections now are mechanical, but plumbing on paper cable is still used.
5. To prevent ingress of moisture	Use of sealant at each end of the termination.
6. To connect to electrical equipment	The lugs used for this purpose will either be of the crimped or mechanical variety. The latter type is growing in popularity because expensive crimping equipment is not required.
7. To insulate exposed connections	Use insulating bushing boots.

The issue of stress control has already been discussed in detail. Some of the other functions of a termination will now be discussed in more detail, apart from earthing and the effects of moisture on the surface of the outer material. These will be covered in the following chapters.

8.2. Management of air

8.2.1. Paper belted terminations

Whilst it is fair to say that paper cable has all but been phased out, there is still a huge quantity of perfectly good cable of this type which will be in service for many

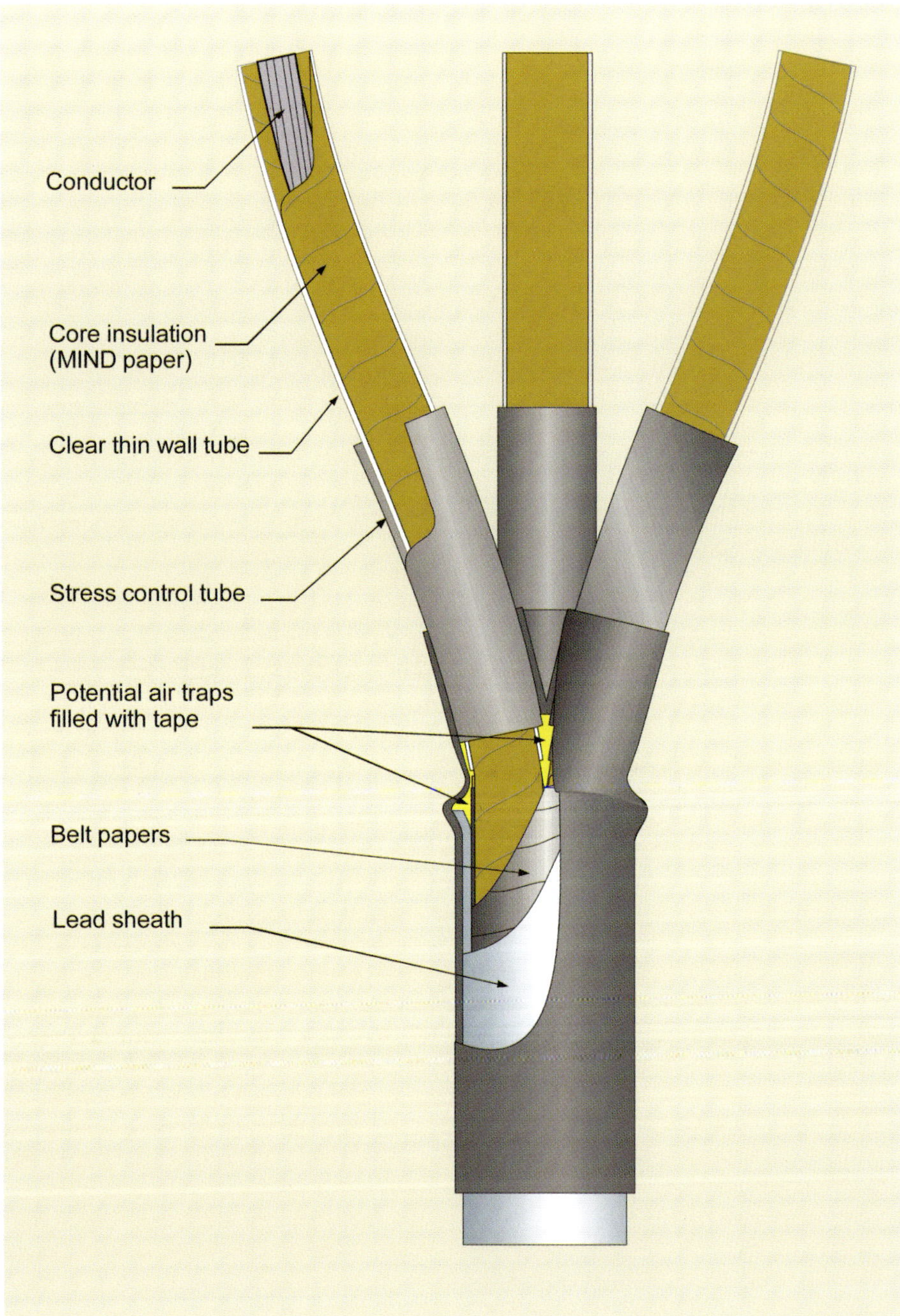

Figure 42
Position of stress control on typical paper belted termination

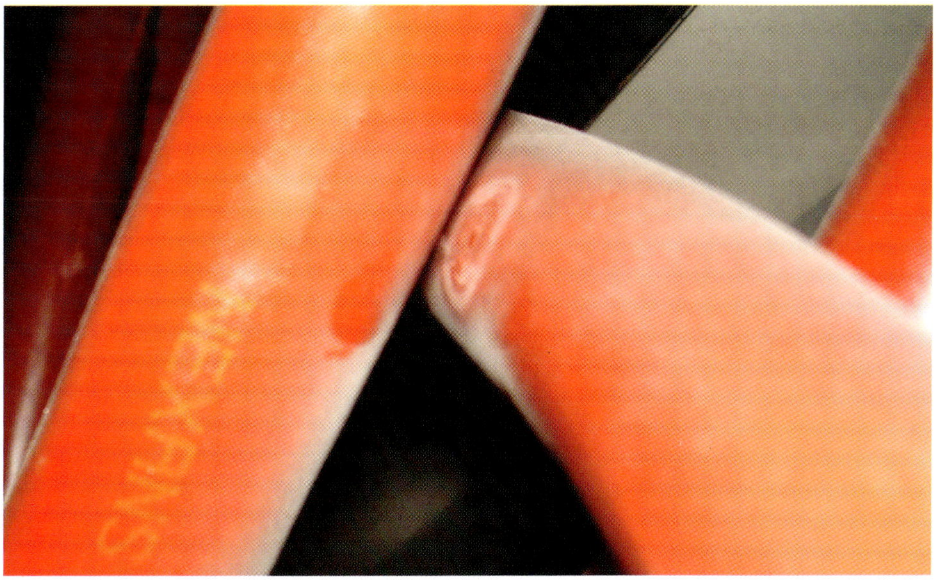

Figure 43
Effect of discharging between crossed cores of a termination

years to come. Therefore a little time will be spent on terminations for paper belted cable which may require terminating or jointing as systems are upgraded. The majority of paper belted terminations supplied require that the stress control system is in contact with the metal sheath by way of a cable breakout. This, of necessity, means that between the top of the stress control and bottom of the lug there is a sizeable amount of unscreened cable as shown in fig. 42.

A similar situation to the above arises with screened cables if the screens are removed close to the cable breakout. The result can be breakdown of the air between cores and this may lead eventually to total breakdown of the termination outer sleeve and cable insulation. The reason for this is that there is a high electrical stress in the region where the cores touch due to the voltage divider effect (see fig 12). The effect of this is shown on an actual termination in fig. 43.

Note the circular discolouration on the right hand core. This is caused by the discharges attacking the surface of the termination and if it had not been found in time would have lead to a complete breakdown. A further indication of discharge activity is shown in fig. 44 where a build up of white deposits are formed.

This discharging can lead to a further problem if the cable box is not completely dry. The electrical discharges break down the air and the insulation to produce ozone which combines with the moist atmosphere to create acids. The acids and ozone

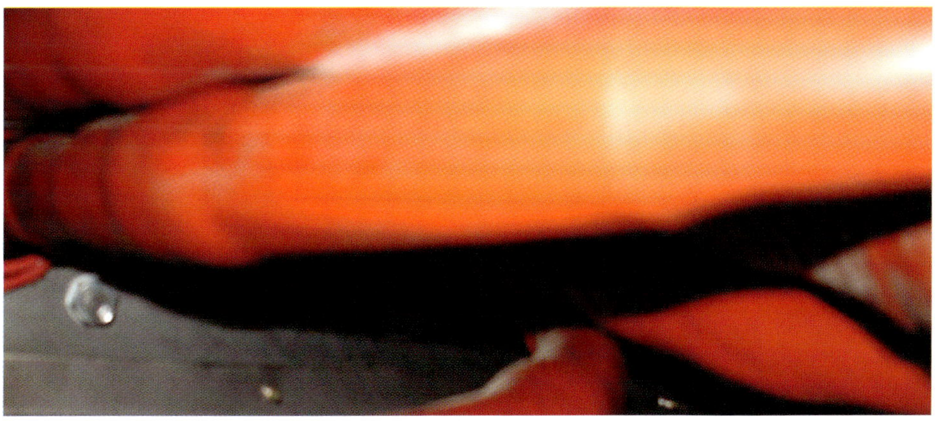

Figure 44
White deposits formed due to discharges

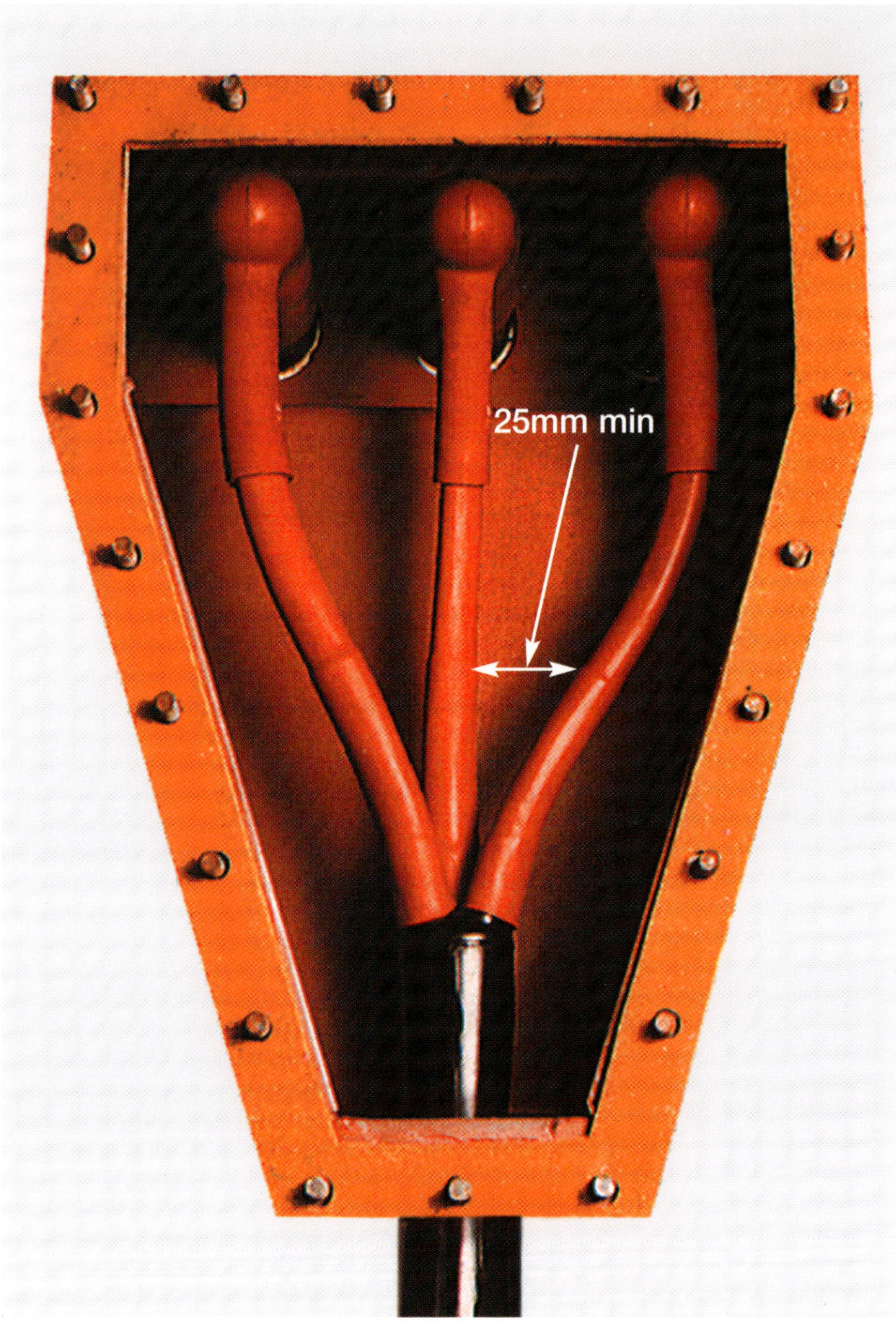

Figure 45
The optimum installation of a paper belted termination

can then attack the cable box to cause corrosion and in addition cause degradation of the bushings, which could initiate tracking and subsequent failure.

The problem with this situation is that pressure testing when the termination is installed will not show the potential fault. It will only eventually be noticeable through audible discharge, or worse, catastrophic failure.

One solution is to not phase out in the cable box, thus enabling a termination as shown in fig. 45 to be achieved with at least 25 mm spacing between cores level with the top of the stress control. The disadvantage of this is that it is a costly and time-consuming exercise since a joint must be installed outside the equipment to carry out the phasing. The cores of cables of less than 120 mm^2 cross section, having a roll in the cable box, can usually be set to achieve the necessary clearance, but beware that the cable will relax over time with the result that discharging may commence at some time after installation.

At this point it may be worth mentioning that when paper cable was generally used for 11 kV, distribution switchgear and transformers were fitted with cable boxes that could be filled with compound. This would exclude the possibility of air being present between the cores.

The higher dielectric strength of the compound would prevent partial discharging, even though the stresses in the crossed area would still be quite high if the cores were close together or touching.

8.2.2. Screened cable terminations

The ideal way of terminating three core cables is to ensure the screens are removed to the dimensions shown in fig. 46 and then it is possible to carry out any cross in the screened area. Unfortunately not all manufacturers show this on their instruction sheets with the result that the stress control is placed too low on the termination. Hence if core crossing is necessary potential trouble awaits.

Figure 46 shows a copper taped screen on XLPE insulated cable but the same dimension should apply for screened paper cable and paper belted 11 kV cable which has been screened with conductive tubing as shown in fig. 33.

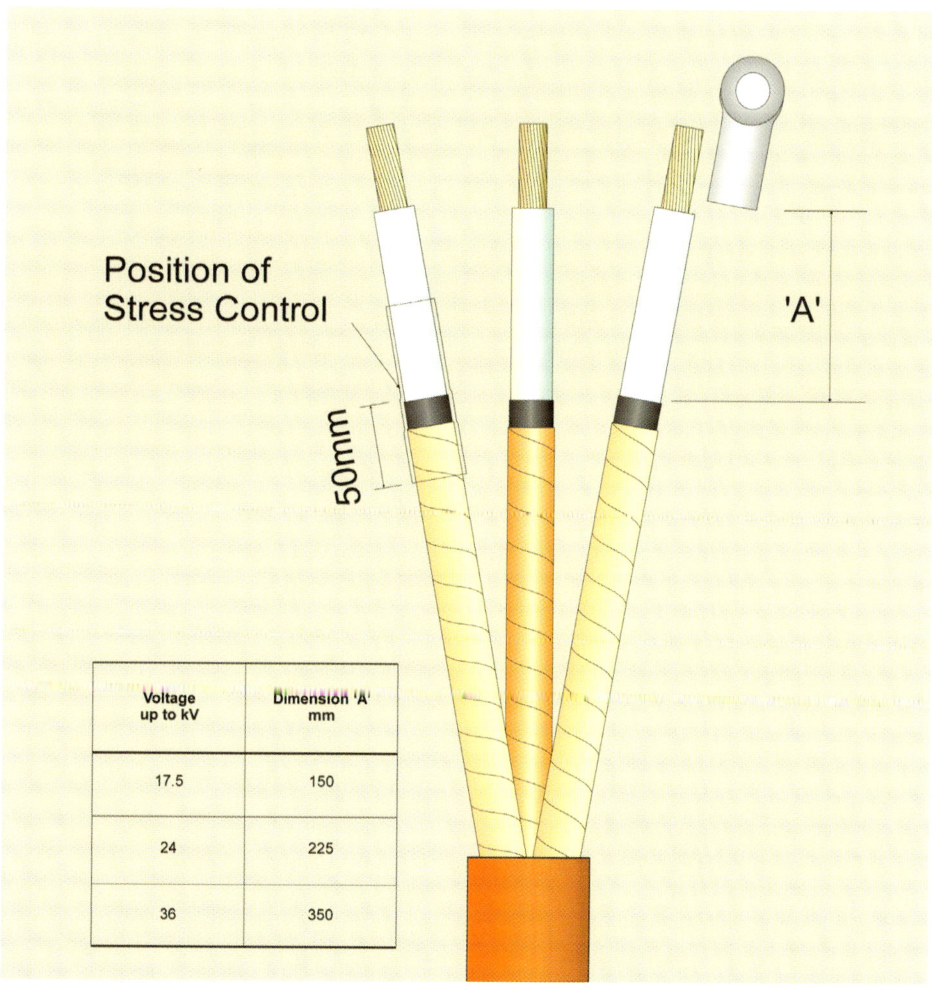

Voltage up to kV	Dimension 'A' mm
17.5	150
24	225
36	350

Figure 46
Stripping dimensions for screened cable

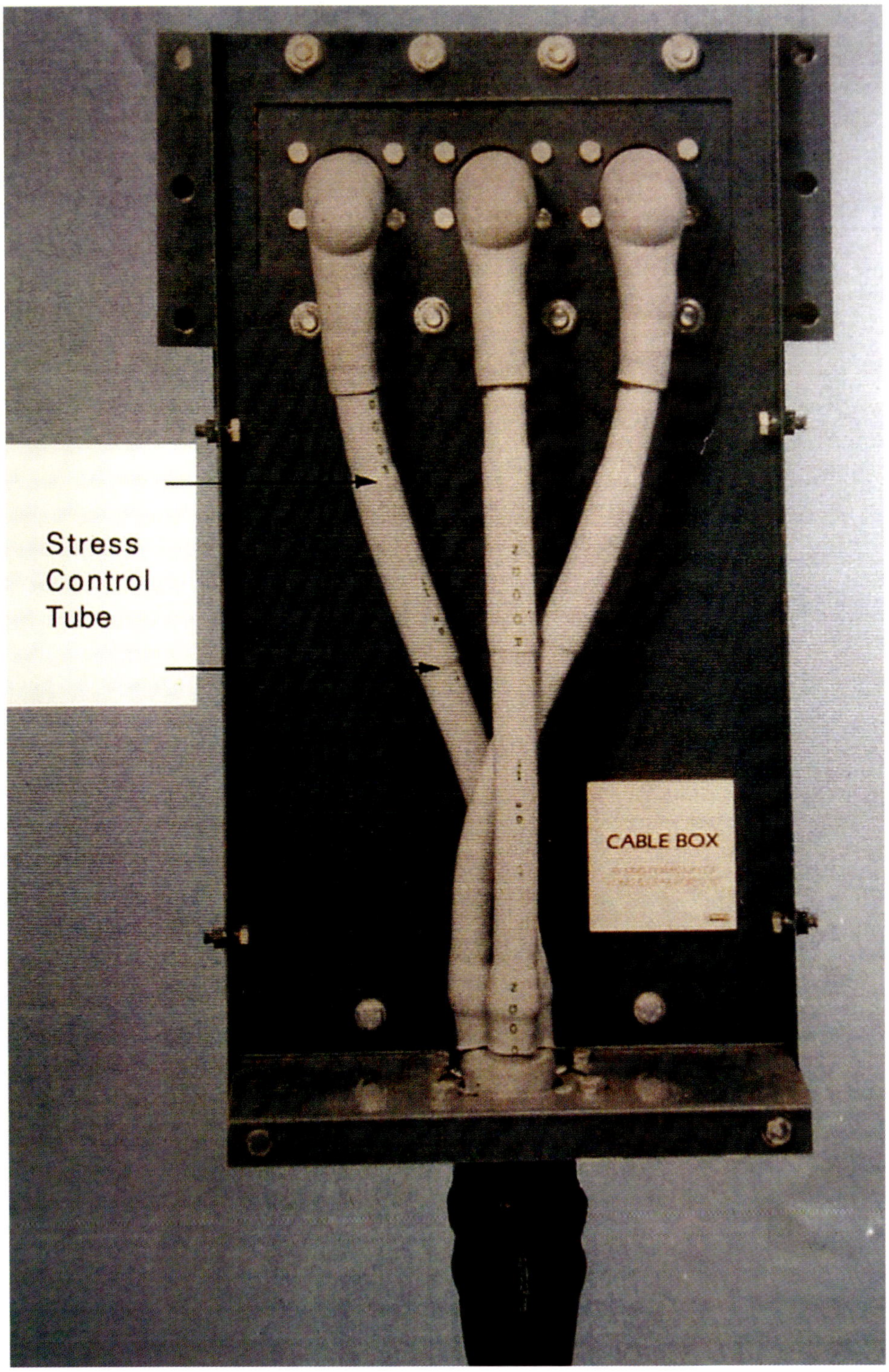

Figure 47
Correctly removed screens to allow safe core crossing

A typical termination on a screened cable with correctly removed screens is shown in fig. 47.

8.2.3. Screened cable terminations – some examples

The following photographs show some terminations installed in the Middle East. The jointers were very good and followed the instructions closely but were ill advised by instruction sheets and old wives' tales. See figs. 48 and 49.

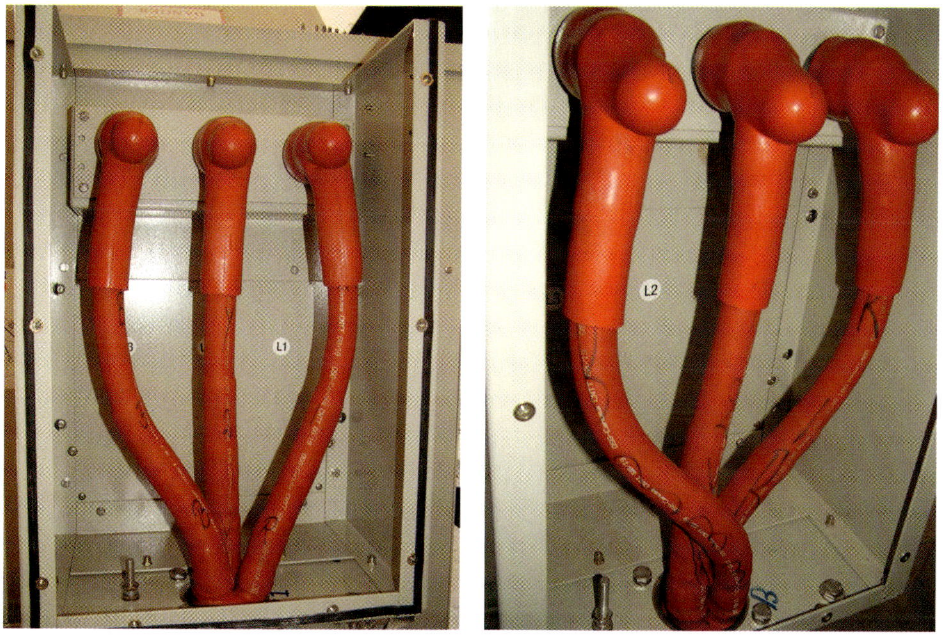

Figure 48
The good
and the bad

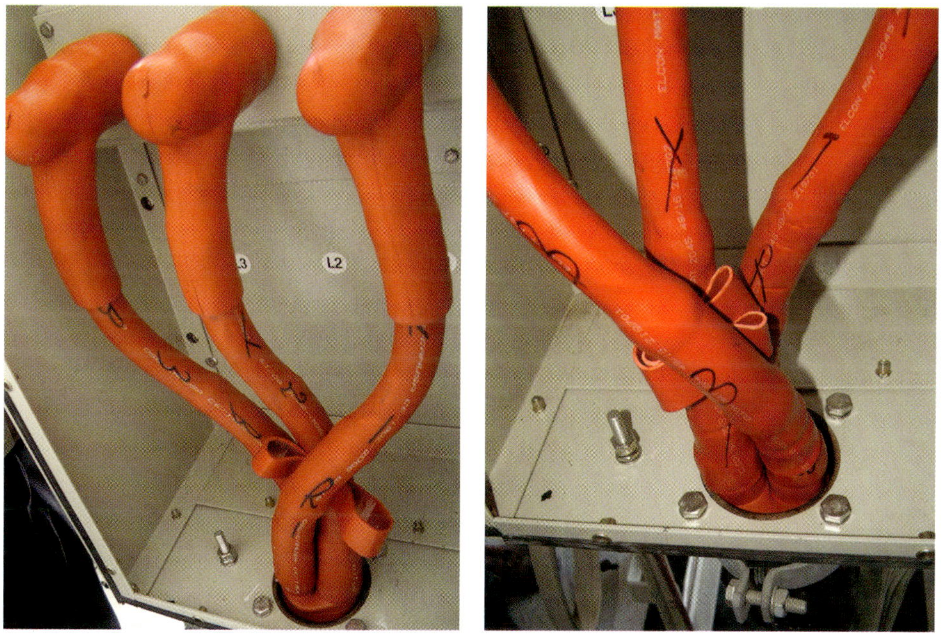

Figure 49
The bad

When looking at figs. 48 and 49 it is clear that one termination will be perfectly satisfactory, whilst the other three potentially will develop problems. Again it must be stressed that the jointers here were very good but the two problems were:

The instructions indicated that the stress control should be situated near the breakout instead of measuring from top down (see fig. 46) hence when core crossing was performed it was in a highly stressed region.

Secondly, the jointers were told to separate the cores at the cross using spare pieces of tubing. This unfortunately, far from correcting the fault, may well exacerbate it.

8.3. Connection to equipment bushings

When a termination is connected to a piece of equipment it may be necessary to insulate the connection. Some equipment is designed for air spacing in which case the termination is connected to the appropriate point and no further insulation is required.

Present day equipment tends to be compact and therefore it is usually necessary to insulate termination connections to ensure that the electrical test requirements are met. There are two basic methods of providing this insulation, namely:

- Heat shrink mouldings
- Push-on mouldings
- Screened separable connectors

The first two methods will now be considered in a little more detail. Screened separable connectors are dealt with in chapter 9.

8.3.1. Heat shrink mouldings

These products have been used for some forty years. They are versatile being able to cope with many bushing designs, and are well proven. There are two basic types depending upon the bushing arrangement and whether it will be at right angles to

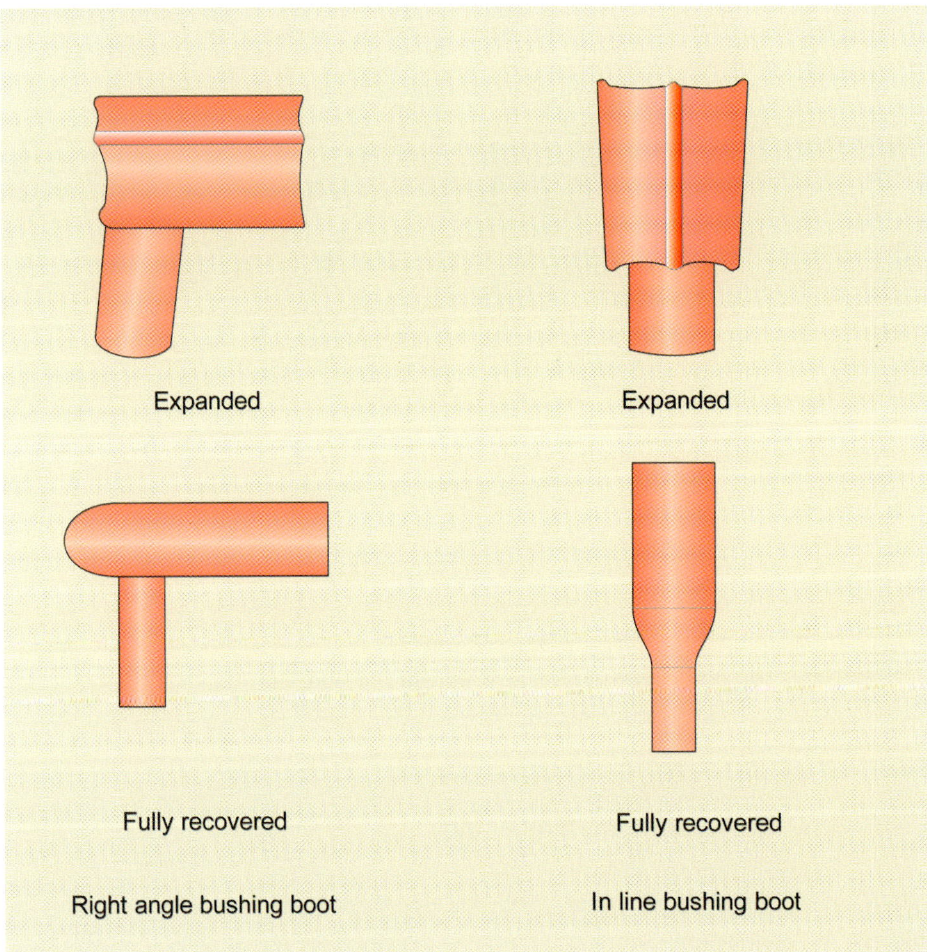

Figure 50 Typical heat shrinkable bushing boots

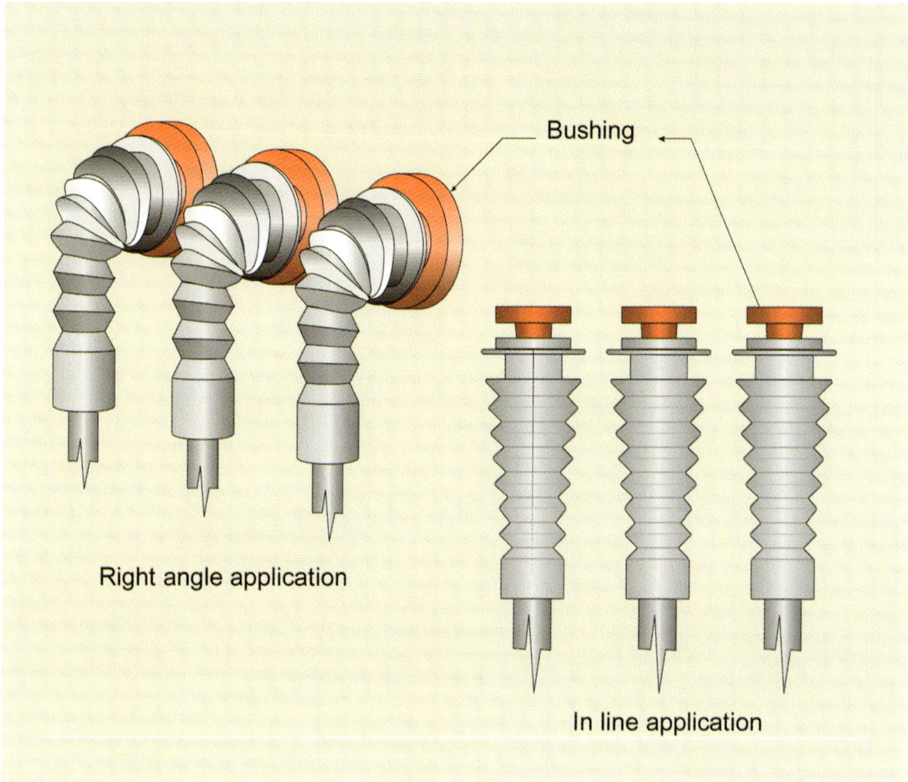

Figure 51
Typical push-on bushing boots

the termination or in line with it. Figure 50 shows the two types in their expanded and recovered states.

Right angle heat shrinkable bushing boots installed in a cable box are shown in figs. 48 and 49. This product is versatile, but more than one size is required to cover the normal range of cable sizes met in practice. The majority of these products are installed on equipment rated up to 17.5 kV with an impulse withstand requirement of 95 kV.

On 36 kV rated equipment a combination of the heat shrinkable boot and rain shed can be produced to meet the 170 kV impulse requirement, but it is necessary to wrap the metalwork around the connection with stress relieving tapes. However, in practice many switchgear and transformer designs are designed for air clearance.

8.3.2. Push on mouldings

This product is becoming widely accepted and has distinct advantages over the heat shrinkable equivalent.

Figure 51 shows a typical product and fig. 52 shows an actual installation on single core cables.

- One basic moulding will cover most bushing and cable sizes although an additional collar may be required to ensure a good seal onto smaller bushings.

- The same product can be used for both right angle and in line applications.

Figure 52
Terminations with push on bushing boots

No heat is required therefore any secondary wiring that may be located nearby will not be damaged during installation.

8.4. Issues with bushing connection insulating systems

Correctly designed and installed systems will not give any problems in service. However, before entering service, one of the most difficult tests to pass is the impulse voltage withstand. This test will show up potential weaknesses in the system and the likely failure points are shown in fig. 53.

This section is concerned only with unscreened bushing shrouds. Screened separable connectors have been produced in conjunction with screened bushings and they work well together. The problems arise with the huge number of designs of unscreened bushings and cable boxes, which have evolved over the years and it has generally fallen on the accessory manufacturer to arrive at suitable solutions.

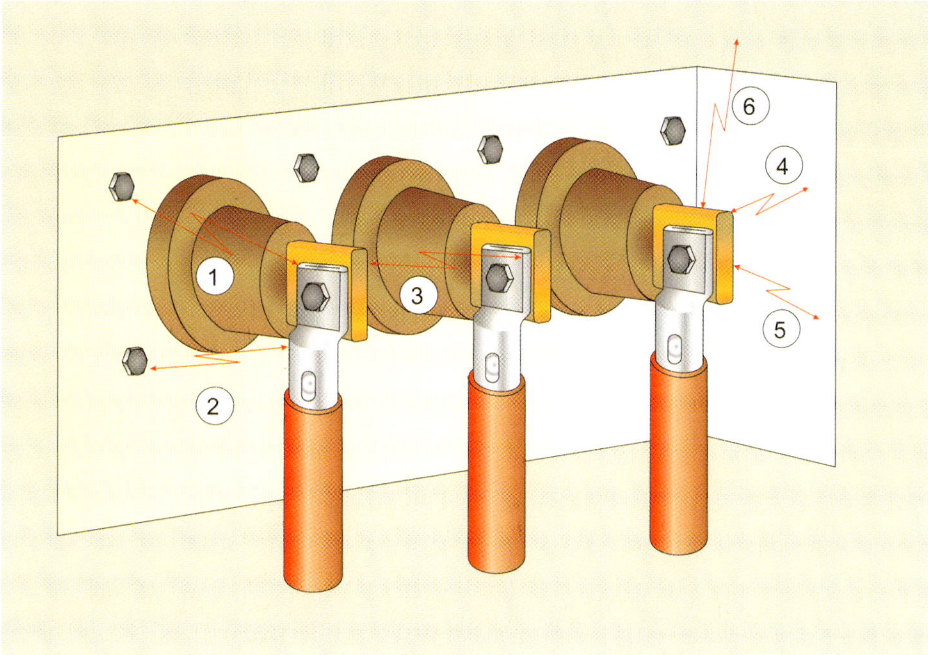

Figure 53
Likely flashover points of bushing/ termination connections

The above diagram shows 6 possible flashover paths when carrying out impulse tests on terminations connected to bushings. The purpose of the bushing boots is to prevent these failures, but unless a tight seal is achieved onto the bushing and the product is produced with sufficient insulation thickness they will not be successful.

In this drawing the connections could be improved if the lugs were positioned on the opposite side of the bushing, but it is a good idea to consider the worst case scenario which could occur in the field. Most of the problems arise because the bushings are extremely short and the earthed metal to the sides and rear is close to live metal.

The various flashover paths shown in fig. 53 will now be considered in a little more detail.

1. From the top of the lug to the back plate. This failure is more likely when the lug palm is larger than the bushing copper bar and the situation is made worse when fixing bolts are used because they not only reduce clearance but also present a sharp edge. The solution is to wrap tape over the metal work.
   In some instances it may be necessary to use stress relieving tape although generally a good insulating tape will suffice.

2. From the lug barrel to the back plate. This is usually solved by the use of insulating tape, although it could be that the shroud insulation is too thin. Extending the termination tube up to the palm, provided it does not prevent good contact with the bushing, can solve the problem.

3. Between phases. This is the result of insufficient separation between bushings. In practice, this type of flashover is less common than the others discussed and it is unlikely to occur when steps have been taken to prevent the other forms of flash over. In the worst case it may be necessary to use phase barriers, but at voltages up to 17.5 kV this should not be required.

4. From the bushing to the side of the cable box. This is the result of the side of the cable box being too close to the bushing. The best solution is for the cable box to be designed to provide adequate clearance, but tape over the metal work will usually suffice. As a last resort, an insulating sheet attached to the side of the cable box would solve the problem.

5. From the bushing to the back of the cable box. This should not be a problem if all the potential weaknesses above have been addressed.

6. From the bushing connection to the top of the cable box. Once again this is not likely to be an issue unless the air clearance is very small.

Flash over occurs on test as a result of the bushing shroud insulation being too thin, the seal onto the bushing being poor, or in some instances, the bushing is simply too small. In the past, unfortunately, equipment manufacturers have designed bushings and cable boxes with little regard for the cable connections. Fortunately, designers are now more aware of the problems and a compromise, which gives a satisfactory solution, is forthcoming.

There are recommended minimum phase separations and clearances to earth suggested by manufacturers of bushing shrouds but these dimensions are only a guide. A system which is well designed may meet the requirements with smaller clearances, whilst a badly designed system with larger clearances may not. The only guaranteed way to ensure a termination/bushing interface meets the specification is to actually test it.

One of the problems facing the end user is that the equipment/termination combination may not have been tested and it may not be feasible to perform such a test in the necessary time scale. However, provided tests have been carried out on similar systems it is reasonable to assume that satisfactory operation will be obtained.

The most likely failure in the field is not due to an over voltage but the result of a poor interfacial seal between the bushing and the shroud. This can lead to moisture ingress along the interface, which will then lead to anaerobic tracking (discussed in a later chapter), and ultimate failure.

In conclusion it can be stated that even though testing of a particular bushing/termination combination has not been carried out this should not be an issue provided that:

- Similar arrangements have been tested.

- The seal onto the bushing is good – this is the most important criteria.

- The bushing shrouds on adjacent cores are not closer than 25 mm for 17.5 kV rated gear and at least 50 mm for 36 kV bushings. If rain sheds are used with the 36 kV system they should also be separated by 50 mm between phases.

- The shrouds are not closer than 15 mm to earthed metal on 17.5 kV and 35 mm on 36 kV equipment.

The above problems can be eliminated by using screened separable connectors to terminate a cable and connect to equipment. These products will be discussed in the following chapter.

8.5. Outdoor terminations

Outdoor terminations are very similar to those used in indoor applications, with some differences as indicated below.

- The length of the termination will normally be longer for a given voltage rating.
- There will probably be rain sheds added to heat shrink terminations to increase the creepage distance.
- Moulded cold shrink and push-on terminations will have additional sheds to increase creepage.
- Three core terminations may have a strengthening component or system around or just above the breakout. This will depend upon the short circuit requirement of the termination. On heat shrink systems a tri-shed is used to hold the cores in place during short circuit. Glass fibre tape wrapped around the cores prior to fitting the termination will serve a similar purpose on polymeric cable.
- Terminations are usually supported using stand off insulators or surge arresters. These both give mechanical support to the termination. The connection to the overhead line or pole mounted transformer is by way of drop wires.

Figure 54
Outdoor termination with stand off insulators

A typical outdoor paper belted termination, which includes a tri-shed to provide addition mechanical strength to restrain the cores under short circuit conditions, and is supported by stand off insulators, is shown in fig. 54.

Surge arresters can be used in place of the stand off insulators to protect the termination, cable and equipment.

Chapter 9

Terminations 2 Separable connectors

9.1. Introduction

An alternative method of terminating a cable is to use separable connectors, also known as screened elbows or tee connectors. These can be deadbreak (cannot connect/disconnect when energised), or loadbreak (loadbreak/fault make). However, this treatise focuses only on the deadbreak type. Theoretically these connectors are the best method of terminating cables as will become apparent later in the book. As with all systems there are pros and cons as listed below.

Pros:

- The connector is fully screened and therefore not affected by adverse environmental conditions.
- The connector consists of two Faraday Cages which eliminates the need for separate stress control and taping of components.
- Tee connectors can be piggy backed to allow more than one cable to be connected to a bushing.
- Dependent on manufacturer, each unit is factory tested for partial discharge and power frequency withstand.
- Products are current rated thus the same connector used on specific bushings may be used for voltages up to 36 kV.
- The products can be used indoor and outdoor.
- They can be utilised to joint cables.

Cons:

- The equipment to which the connector is fitted must have the correct bushing interface, unlike with heat shrink and cold shrink terminations.

- The connectors are best suited to single core cables although three core cables can be modified to suit.
- The product tends to be more expensive than the equivalent heat shrink and cold shrink termination.

9.2. A Screened Separable Connector

Figure 55 shows a typical screened elbow rated at up to 24 kV and 250 Amps. The connection to the bushing equipment is a plug-in arrangement. It is noted that there is a conductive moulding on the right hand side of the elbow and this is to provide a low voltage take off to indicate if the circuit is live. This insert provides a capacitive voltage divider and can be used in conjunction with a light to show the circuit is live. Note the earth lead attached to the conductive outer layer of the connector. This is essential to ensure the outer surface potential is defined to provide the geometric stress control of the system.

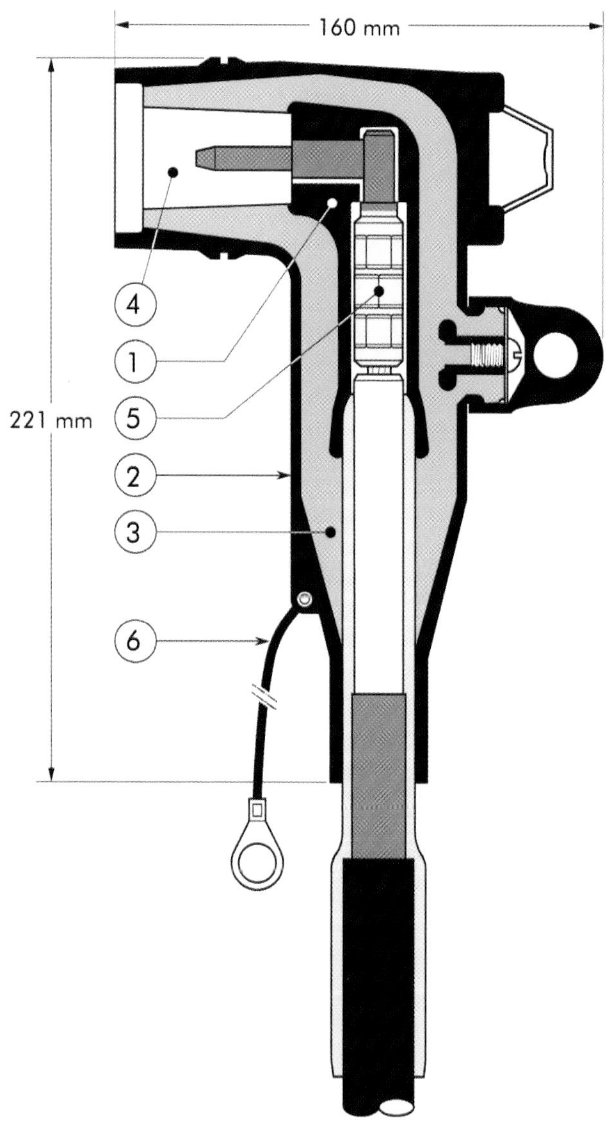

1. Conductive EPDM insert.
2. Conductive EPDM jacket.
3. Insulating EPDM layer moulded between the inert and the jacket.
4. Type A – 250 A interface as described by CENELEC EN 50180 and 50181.
5. Conductor connector.
6. Earthing lead.

Figure 55
A 250 A
Screened Elbow

9.3. Bushings

Before other types of screened connectors are considered the following looks at the bushing designs which are produced in accordance with CENELEC EN 50180/ 50181 and shown in fig. 56.

Rated Voltage kV	Rated Current A	D_1 mm	D_2 mm	D_3 mm	L_1 mm	L_2 mm	Contact Type	Interface
12 – 24	250	31	32.5	48.5	48	9	Sliding	A
12 – 36	250/400	46	56	70	90	11	Sliding	B1/B2
12 – 36	630/1250	46	56	70	90	11	Bolted	C1/C2
12 – 24	800/1250	39.9	52.1	76.2	81	14.2	Bolted	D1/D2
36	800/1250	39.9	61.5	76.2	103.7	14.8	Bolted	E1/E2

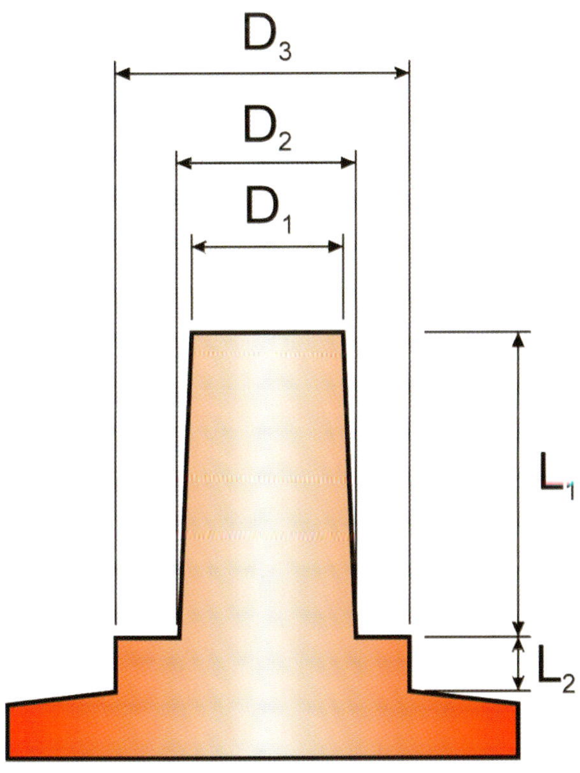

Figure 56
Bushing dimensions to accept separable connectors

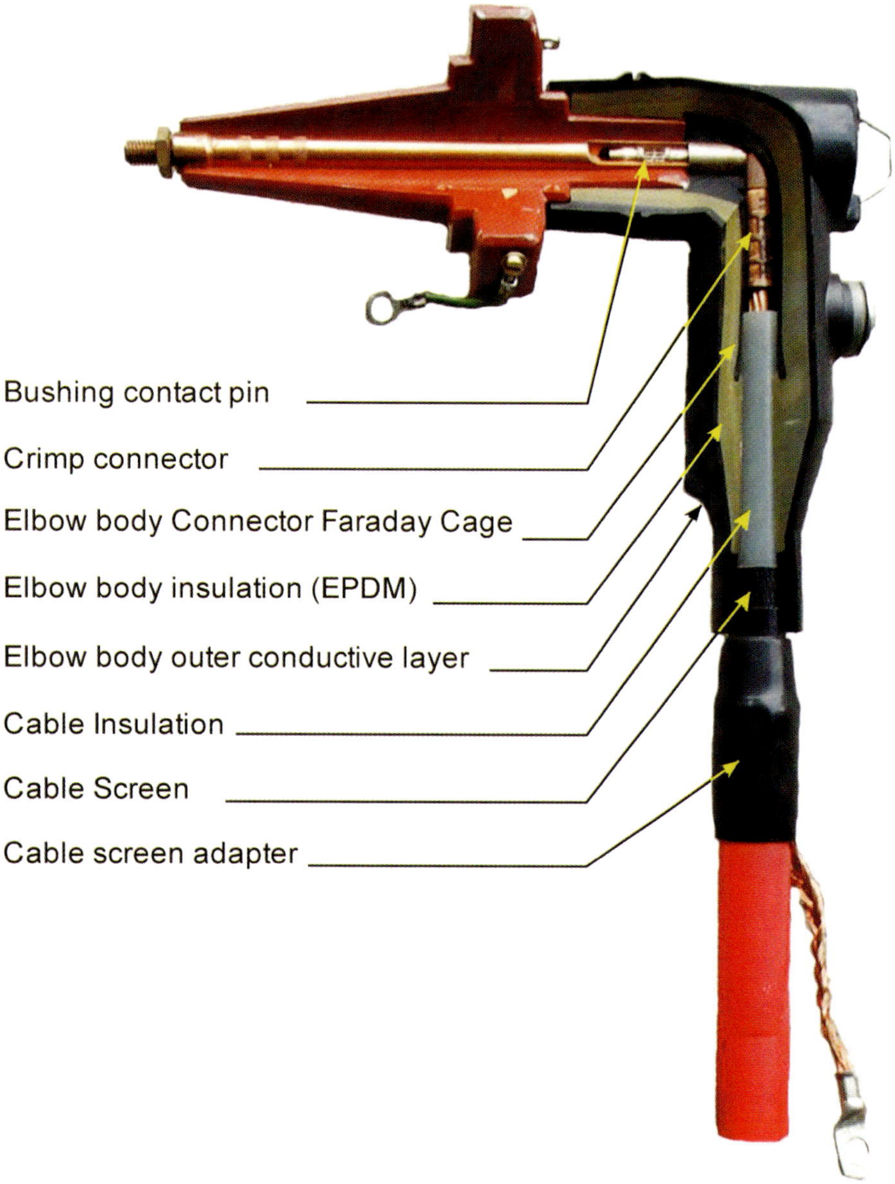

Figure 57
Cross section of bushing and elbow

Figure 57 is a cross section of a bushing and elbow connector. It will be seen that there are very few components required and consist of:

Main body

Crimp connector

Bushing contact pin (which screws into the crimp connector)

9.4. Cable reducers

It has already been stated that the main body of the connector does not readily lend itself to range taking and therefore to reduce the number of connectors required cable reducers are used to increase the diameter over the cable insulation. Figure 58 shows a reducer fitted. Note that the top of the reducer is within the Faraday cage of the connector body thus there is no stress introduced by the reducer.

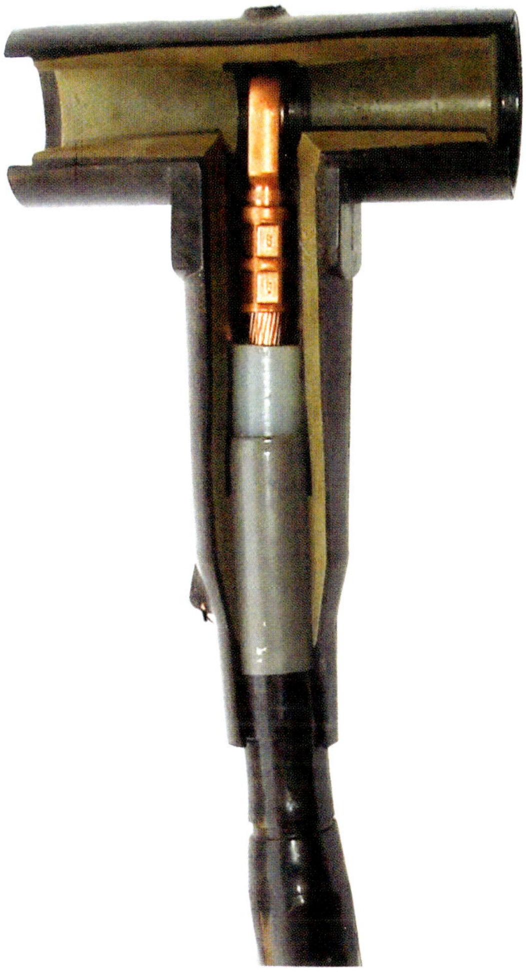

Figure 58
Tee connector
with reducer

9.5. Connecting to a bushing

The process of connecting to a bushing is shown in the series of diagrams in figs. 59, 60 and 61.

Lubrication using silicone grease is necessary to enable the connector to be pushed into place. The clamping screw tightening torque is 50 Nm.

Once fitted to the bushing if only one cable is to be installed per phase then an insulating plug is attached as shown in fig. 61. The torque applied is again 50 Nm.

When removed the voltage detection cap gives access for a check to be carried out to ascertain whether the circuit is live. Note the earth attached to the outer body of the connector to ensure a fully earthed outer surface. The connector should not be allowed to support the weight of the cable and a suitable clamp should be used as shown.

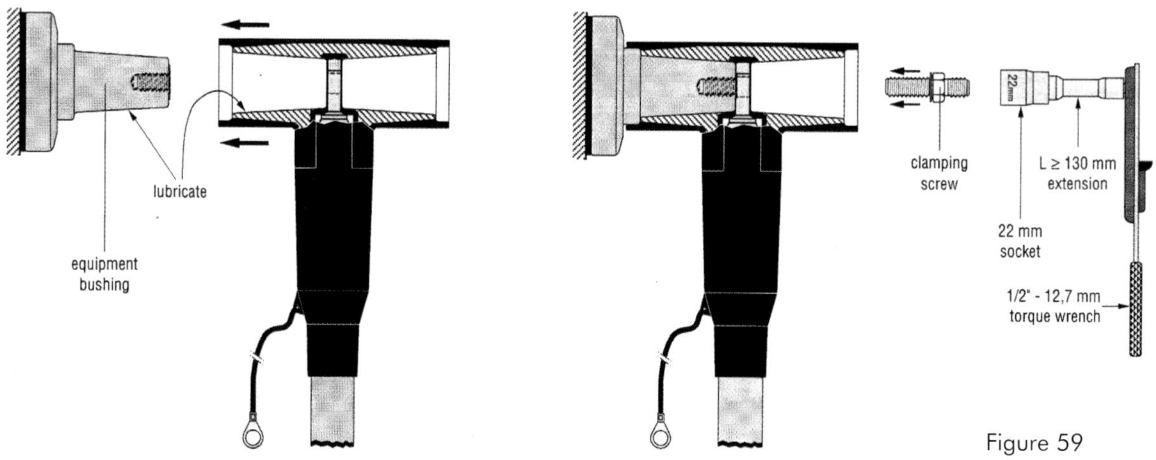

Figure 59
Installing the connector onto the bushing – steps 1 & 2

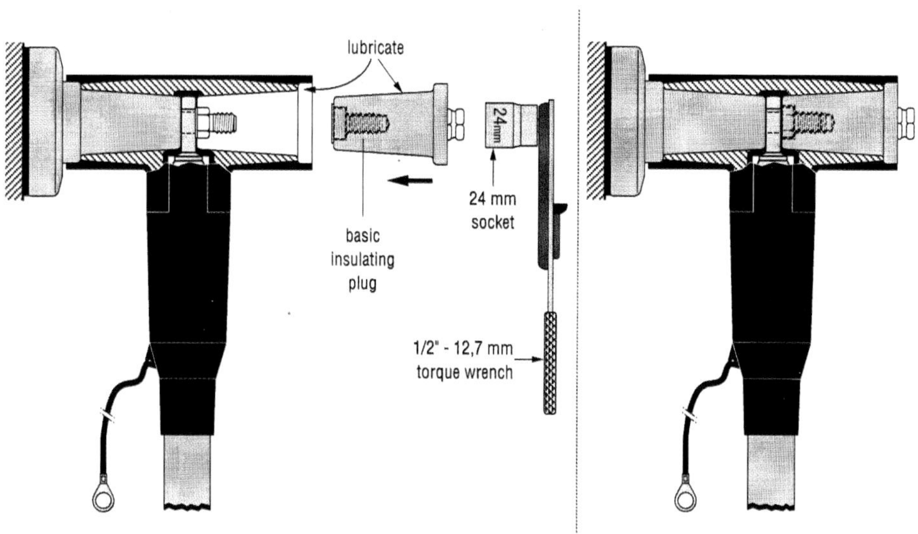

Figure 60
Installing the insulating plug to the connector

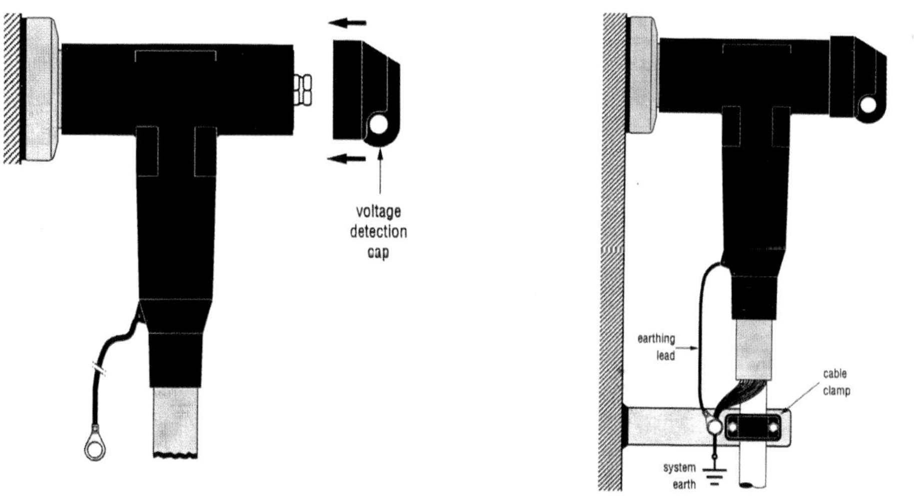

Figure 61
Completing the installation

9.6. Connecting more than one cable per phase

By using a connecting plug, having connected the first connector to the bushing, further connectors can be added as shown in fig. 62 below:

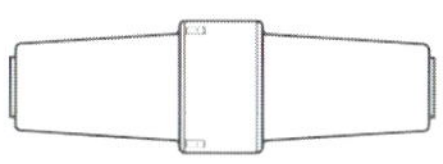

Figure 62 – Connecting two cables per phase

Chapter 10

Terminations 3 Earthing

10.1. Introduction

Cables are earthed at the termination. On single core cables there are various options, but great care should be taken. At first sight it may appear that only one end of the cable need be earthed via the termination. If the cable run is very short (a few metres), for example, where there is a loop connecting a transformer to a ring main unit then this will be satisfactory. However, only earthing one end will give rise to an induced voltage at the other end of the cable earth. Should both ends be earthed, then this eliminates the possibility of induced voltages, but gives rise to induced current in the cable earth. This current can be of a similar value to the current in the conductor if steps are not taken to prevent it. Any significant cable run using three single core cables should adopt the cross bonding technique to reduce circulating earth currents. This will be discussed in more detail in the section on joints. In this chapter, only the method of earthing a termination will be considered.

The function of an earth on a termination is to provide an earth connection to the cable screens, metal sheaths and armours. The earth connection should be capable of carrying any circulating currents without overheating, in addition to the fault current of the system. When designing an earth connection the following points should be taken into consideration:

- Do not use materials which will cause corrosion when in contact with each other.

- Do use tinned copper braid wrapped around armours, if appropriate, to give good contact.

- Ensure that water cannot enter the system and cause corrosion. A solder block on copper braid is essential to prevent this.

- Always use under armour supports on aluminium armour (single core) and steel wire armour (three core) cables to prevent damage to the cable if the system requires direct contact between the armours and the earth lead.

- Ensure that the earth lead is capable of carrying both circulating and short circuit currents without distress.

- Use good quality components.

- Never use ferrous metal glands and support rings on single core cables.

10.2. Outdoor terminations

10.2.1. Single core cable with copper wires or copper tape screen

Most single core cables are constructed with copper wire or copper tape screens and earthing is quite simple. Figure 63 shows a copper wire construction and in this case the wires are bent back along the cable after the outer sheath has been removed and a lug crimped to the wires. The lug can then be connected to an earth point on the support structure. It should be borne in mind that the current carrying capacity in the wires once they are bent away from the cable core is reduced, but this does not normally cause a problem.

Figure 63
Earthing single core cable with copper wires

Figure 64 shows a method of earthing a copper tape screen. This involves simply connecting a copper braid to the screen as shown using a constant force spring. It is possible to make a soldered connection, but the use of the spring is simpler and quicker and tends to be the preferred option. The braid is then connected to an earth point. Note the solder block in the braid.

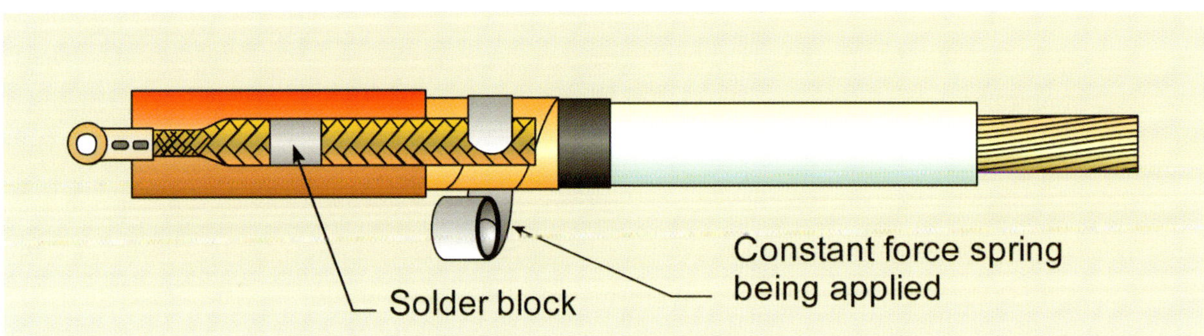

Figure 64
Earthing single core cable with copper tape screen

When the cable is armoured, it is usual to use an aluminium cable gland as discussed below.

10.2.2. Single core polymeric cable with aluminium armour

Single core cables with aluminium armour require an earth connection to be made to the screen and the armours as shown in fig. 65. It should be noted that a support ring of aluminium is fitted under the armours to prevent damage to the cable whist fully tightening the clamps. In addition, by having a rigid base for the armours continuous good contact is maintained between all components. Without a support ring, contact could reduce in time due to the thermal cycling of the cable.

An under armour support of ferrous metal should not be used under any circumstances because circulating currents will be induced, causing potentially damaging overheating. Whilst worm drive clamps are shown in fig. 65, constant force springs can be used.

Finally the exposed armours and earthing system must be covered to provide corrosion protection.

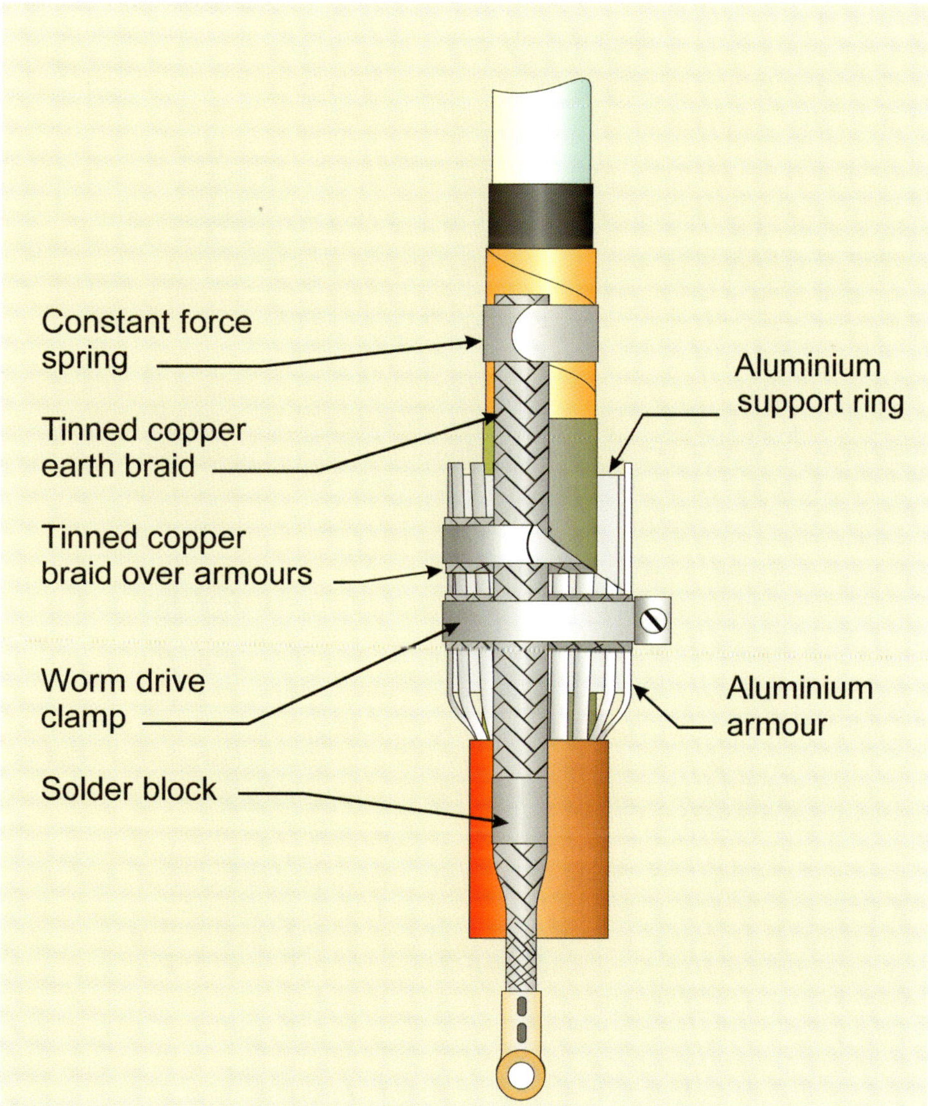

Figure 65
Earthing of single core aluminium tape armoured cable

10.2.3. Three core polymeric cable with steel wire armour

A similar system to that shown in fig. 65 can be used for this cable. The system is shown in greater detail in fig. 66. The under armour support rings used in this instance will probably be of steel as there will be no induced currents created by a three core cable.

When earthing a three core cable the individual core screens must be connected to the metal sheath and/or armours. It is more usual now to use a separate earth braid for each core as shown in fig. 66, although one braid wrapped around all three cores and soldered to each screen can be used.

Notice the use of the tinned copper braid wrapped around the armours prior to connecting the screen earths with the main earth lead. This ensures good electrical contact between all components in the system.

The main earth lead will contain a solder block to prevent water ingress and tape sealant is applied under and over the block to give a complete seal. When the cable termination is fitted the three core breakout body will overlap the corrosion protection sleeve.

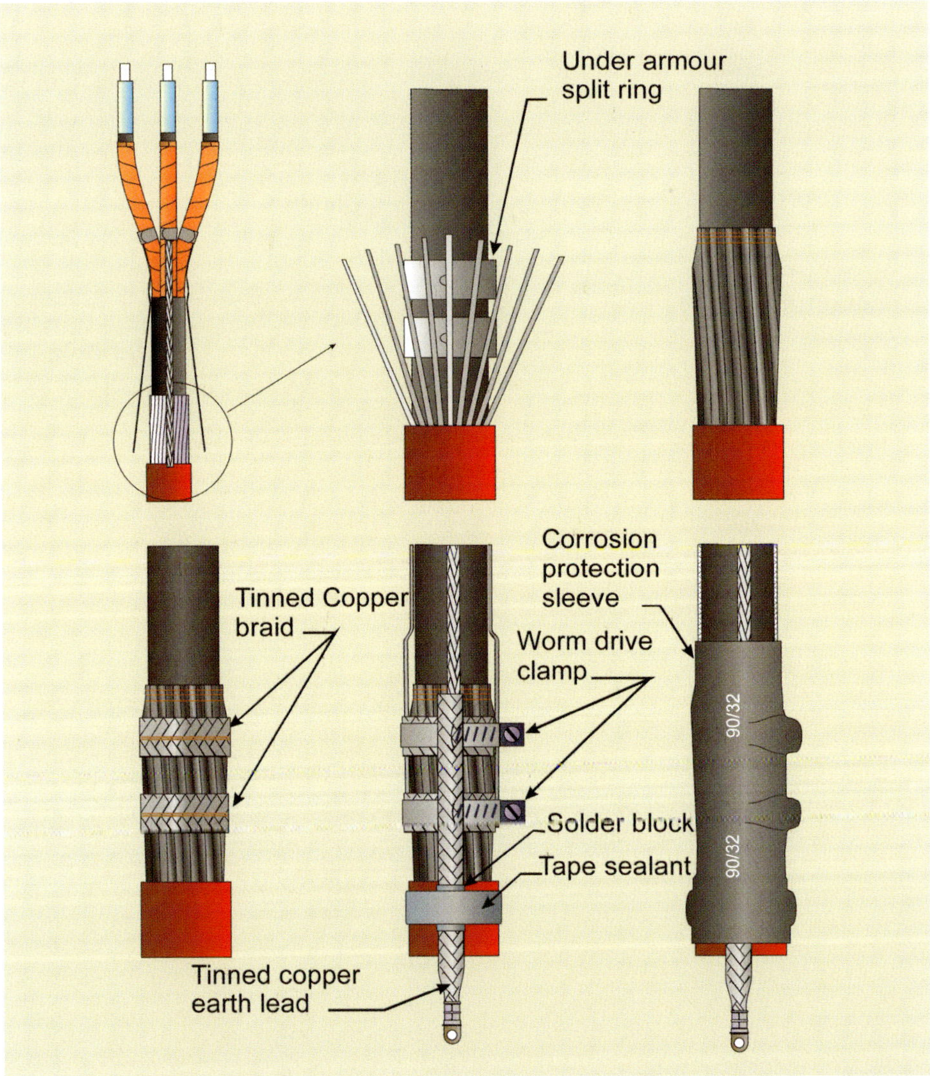

Figure 66
Earthing system for steel wire armour polymeric cable

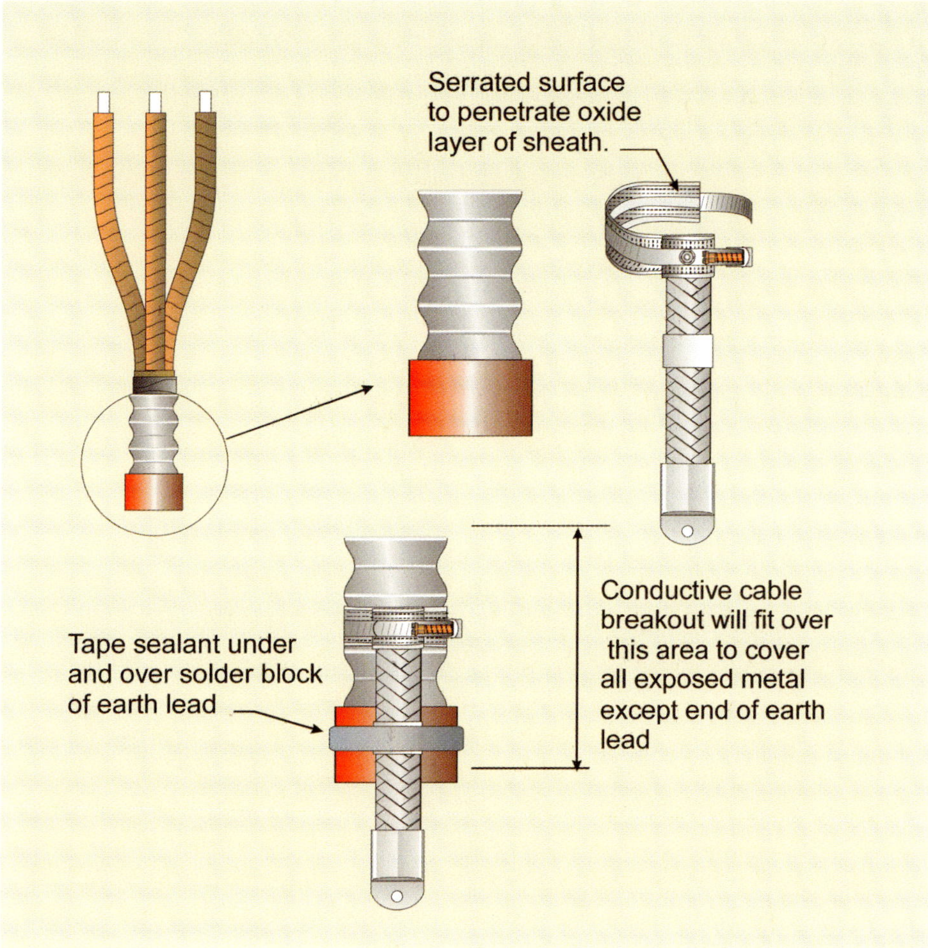

Figure 67
Mechanical earth for corrugated aluminium sheathed cable

10.2.4. Three core paper cable with corrugated aluminium sheath

The use of a solder wipe is a satisfactory method of earthing corrugated aluminium sheathed paper cables, but alternative mechanical methods are now available which require less skill to install than the solder wipe. Furthermore, there is no danger of excess compound being forced out of the papers due to the heat required to perform a satisfactory wipe. Such a system is shown in fig. 67.

On this system the earth lead, serrated edged contact and the worm drive clamp is presented as a single unit for ease of connection. The components can be applied separately and a constant force spring in place of a worm drive clamp could be used.

10.2.5. Paper cable with lead sheath with steel wire armours

As with the corrugated aluminium sheath a solder wipe can be used. A mechanical system as shown in fig. 68, is now the accepted method. Note that constant force springs are used to hold the earth braid onto the lead. The use of worm drive clamps is not advised at this point because over tightening can damage the lead and insulation underneath. Worm drive clamps should, however, be used over the armours to ensure good contact between the armours and earth lead.

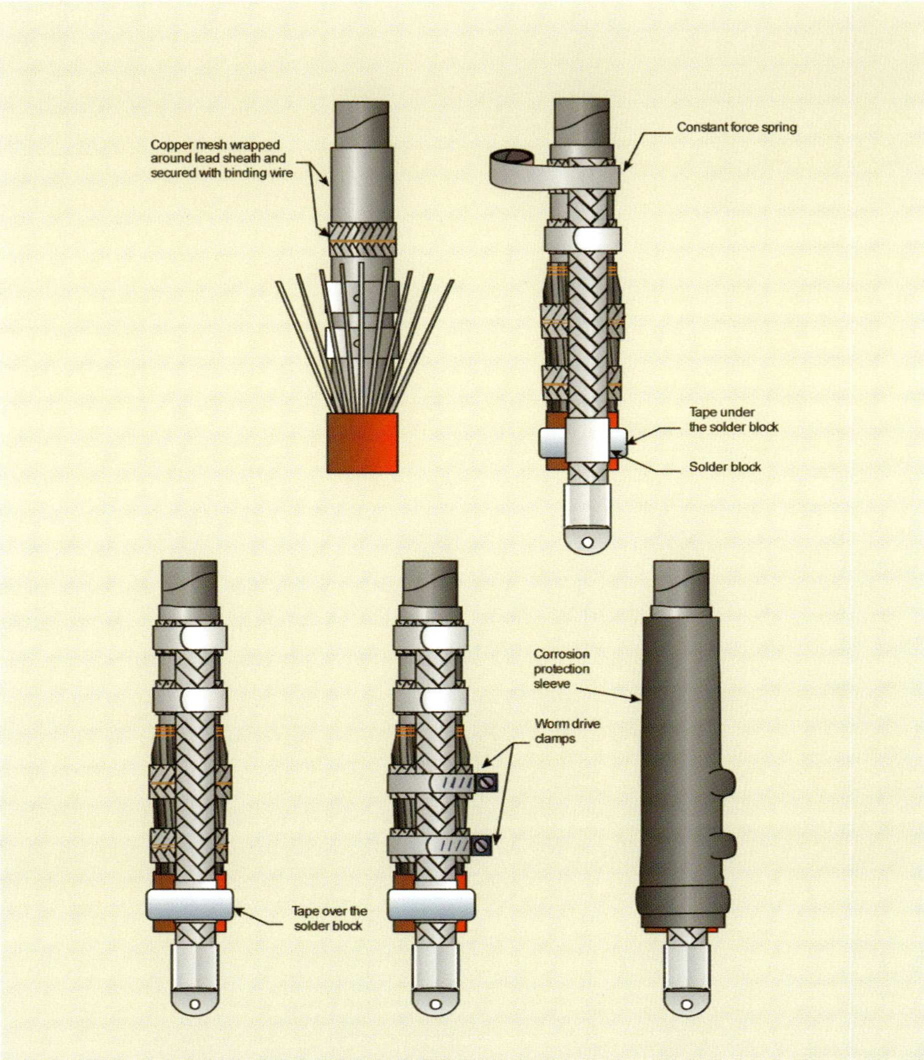

Figure 68
A mechanical earthing system for paper lead cable

10.2.6. Three core paper cable with steel tape armour

This cable construction is less common than steel wire armours, but the earthing is almost identical to that shown in fig. 68 except that under armour rings are not required.

10.3. Indoor terminations

Indoor terminations are often installed inside a cable box and the point of entry of the cable into the box will generally be by way of a gland. For dry type terminations, one of two general types will be employed. For armoured cable a metal gland will be used and the armours will be attached to the body of the gland. Single core cables will require an aluminium gland to prevent induced circulating currents, whilst three core cables will normally use a galvanised mild steel gland. Non armoured cables, including corrugated aluminium sheathed paper cables, can utilise an insulating heat shrinkable gland which provides a seal for the cable as it enters the cable box. The two types are shown in fig. 69. Compression glands are also an option and will be considered in detail later, but tend to be more expensive.

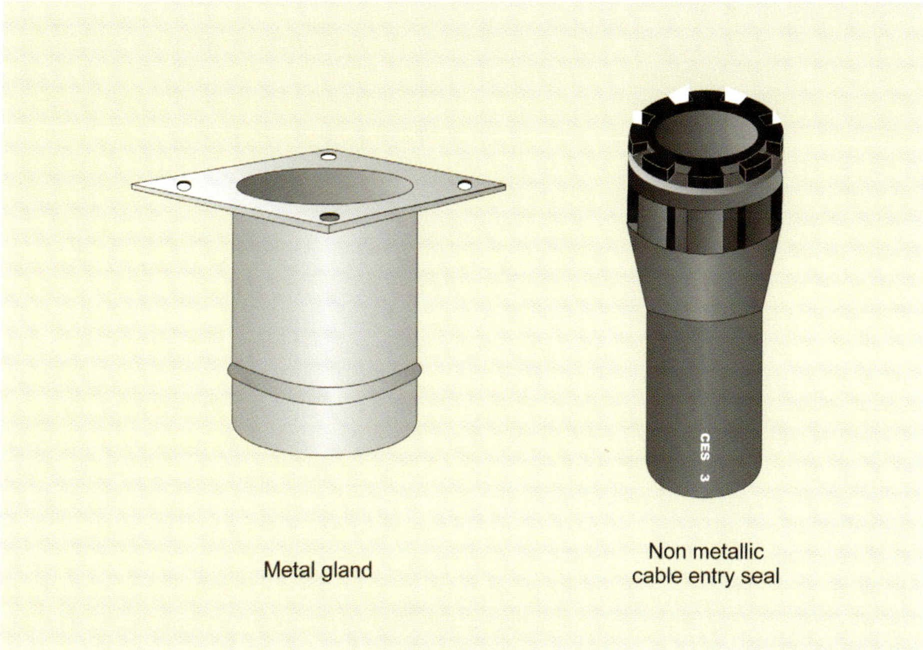

Figure 69
Glands used for indoor terminations

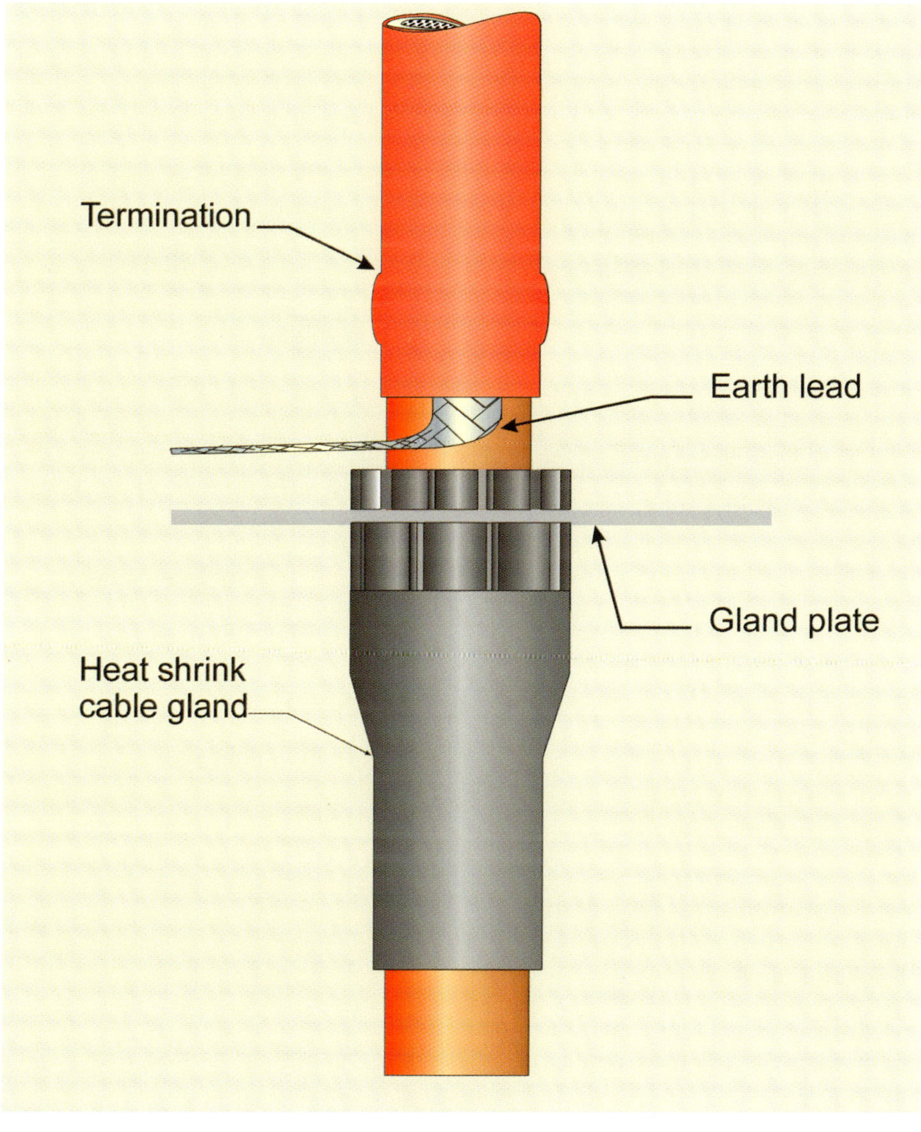

Figure 70
Cable gland system for single core unarmoured cable

10.3.1. Single core unarmoured polymeric cable

Earthing of this type of cable is the same as for outdoor terminations, with an earth connection being made to an earth point in the equipment or cable box. The entry into the equipment will be in the form of gland if a cable box is fitted. Should there not be a cable box a cable clamp fitted below the termination will be required. Under no circumstances should the clamp be fitted on any part of the cable termination as this will lead to discharging and eventual failure. A typical arrangement, employing a heat shrinkable gland, is shown in fig. 70.

10.3.2. Single core cable with tape screen and armours

When earthing armoured single core cable it is necessary to connect the armours to earth in such a way that there is no reduction in the current carrying capacity of the armours. In order to achieve this, an aluminium gland to which the armours are attached can be used as shown in fig. 71. Note that the tape screen of the cable is earthed as shown in fig. 64. There will usually be a stud on the gland in order that an earth lead can be attached to the equipment earth. In order to prevent corrosion, a heat shrinkable or similar sleeve is fitted over the gland and onto the cable sheath. This has not been shown in fig. 71 for the purpose of clarity.

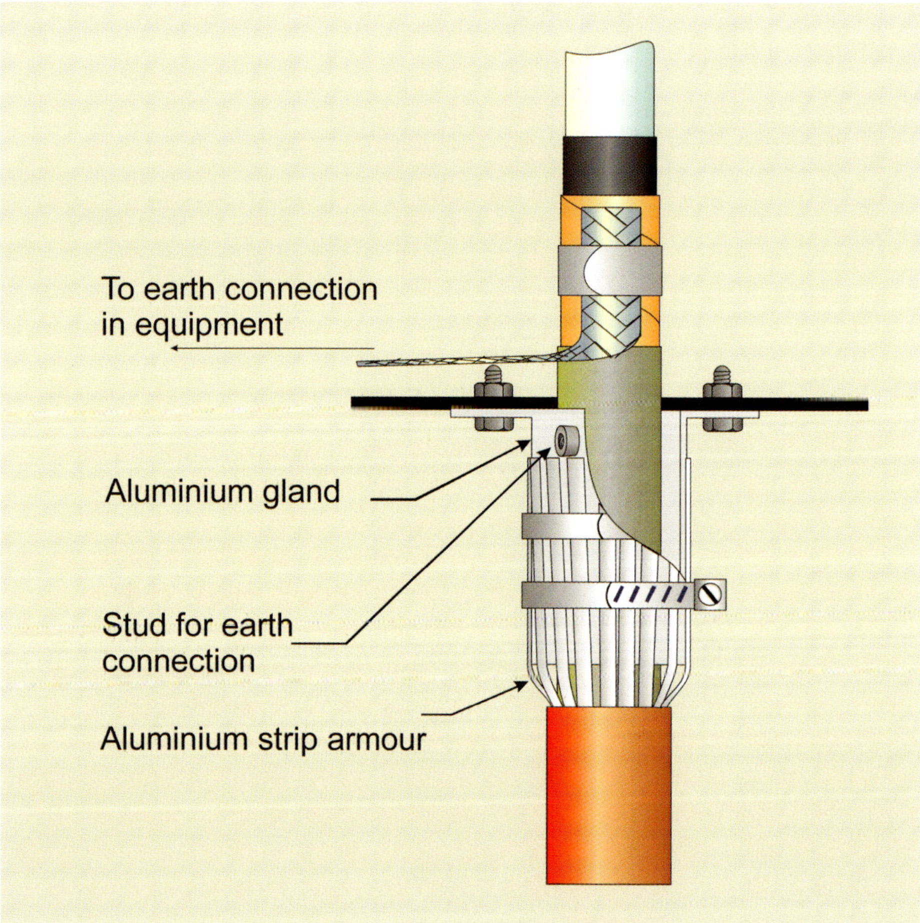

Figure 71
Earth gland system for single core aluminium tape armoured cable

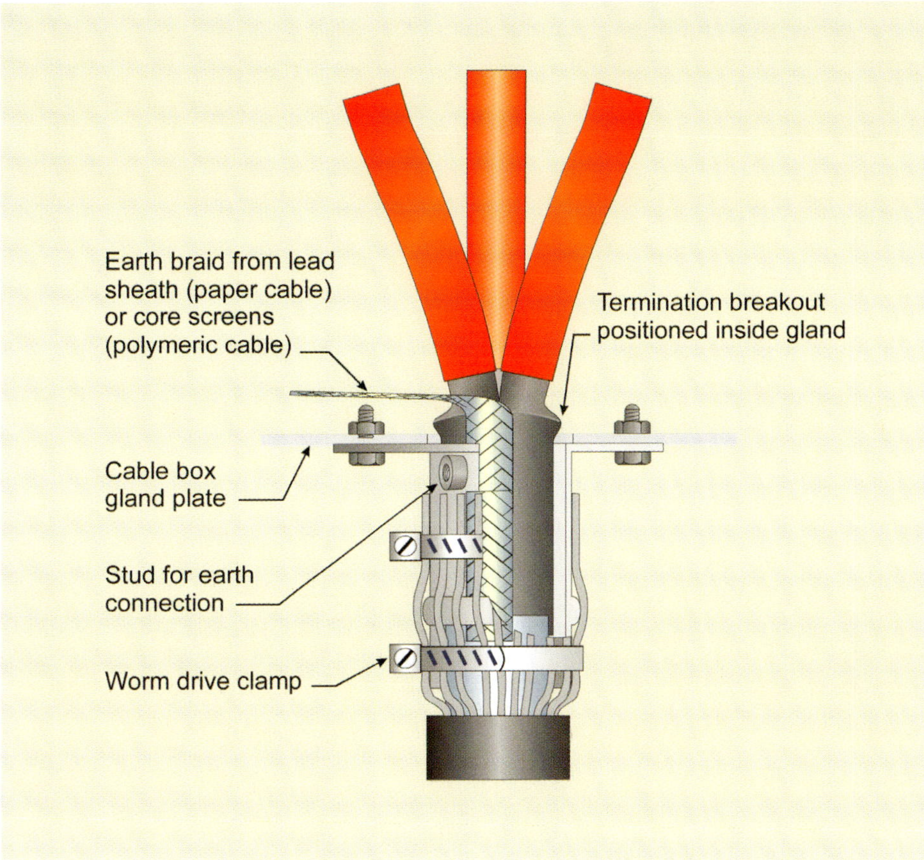

Figure 72
Arrangement for connecting armours to gland

10.3.3. Three core paper cable with corrugated aluminium sheath

Earthing this type of paper cable was traditionally carried out by plumbing the sheath to a brass gland. However, as this procedure requires considerable heat, which can drive excessive cable compound from the end of the cable, mechanical systems are now commonplace. This reduction in the amount of compound in the termination area has been blamed for failures in the past due to the papers in the area around the wipe being electrically weakened.

The more usual method of earthing this type of cable is to employ the mechanical system shown in fig. 67 and then use a cable entry seal similar to that shown in fig. 70. This arrangement will seal the cable entry into a cable box and also give the cable some additional support.

10.3.4. Three core cable with steel wire armours

The system for earthing steel wire armours can be the same for both paper and polymeric cable. A gland similar to that shown in fig. 69 is used and the armours are clamped to the body of the gland. An advantage of this system is that if the cable box is quite short, the breakout of the termination can be positioned inside the gland, which can gain about 100 mm extra for setting the termination cores.

This system is shown in greater detail in fig. 72. The gland in this instance will be of mild steel, which has been galvanised or similarly plated. A sleeve will then be fitted over the armours and clamps to prevent corrosion of the system.

10.4. Compression glands

The previous sections have explained the use of glands shown in fig. 69 but there is a wide range of more sophisticated glands available and these will now be addressed.

Cable glands are used to seal and support cable entry into enclosures and perform one or more of the following functions:

- Environmental protection.
- Cable support.
- Earth continuity.

10.4.1. Ingress protection – IP ratings

The gland will seal the point of entry of the cable to the enclosure. The degree of sealing will depend on individual requirements and is specified as an IP (ingress protection) rating and should be equal to that of the enclosure.

An IP rating is expressed as a 2 digit number. The first digit refers to the degree of protection to a person and the enclosure to foreign bodies. The second refers to the degree of protection afforded to the equipment within the enclosure.

The first digit

0 No special protection
1 Protection form objects > 50 mm diameter including hands
2 Protection against fingers and objects <80 mm long and <12 mm diameter
3 Protection from tools, wires etc > 1 mm diameter
4 Protection from solid objects > 1 mm diameter
5 Protection from amount of dust that would interfere with the operation of equipment
6 Dust tight

The second digit

0 No special protection
1 Protection from dripping water
2 Protection from vertically dripping water
3 Protection from sprayed water
4 Protection from splashed water
5 Protection from water sprayed by a nozzle
6 Protection against heavy sea or powerful jets of water
7 Protection against temporary immersion in water – up to 1 m depth
8 Protection against prolonged immersion in water – beyond 1 m depth

- Thus for a rating of IP56 the gland would ensure operation of the equipment and not allow water penetration provided the gland was not subjected to immersion in water.

- This rating would meet most requirements in industrial environments.

- Occasionally a third digit will be included to specify mechanical performance.

10.4.2. Gland seals

Compression

This comprises an elastomeric ring with a V groove that will close and create a seal onto the cable. See figure 73. The action of tightening of the seal is performed simultaneously with the armour locking action, which may lead to excessive force being applied to the inner sheath of the cable. This can only be avoided by first removing the seal and performing the armour lock and then completing the seal.

Displacement

The displacement seal is gradually pushed down a taper until it makes a seal onto the cable. (Fig. 74) With this arrangement the gradual force applied on the inner sheath of the cable can be controlled independently of the armour locking action, and the user can test the gas seal. In this way the over tightening of the seal can be avoided, without the need to remove any components and/or perform a double make-off.

Diaphragm

This seal overcomes the possibility of damage to the inner heading due to excessive tightening and consists of an elastomeric membrane attached to the armour cone. (Fig. 75)

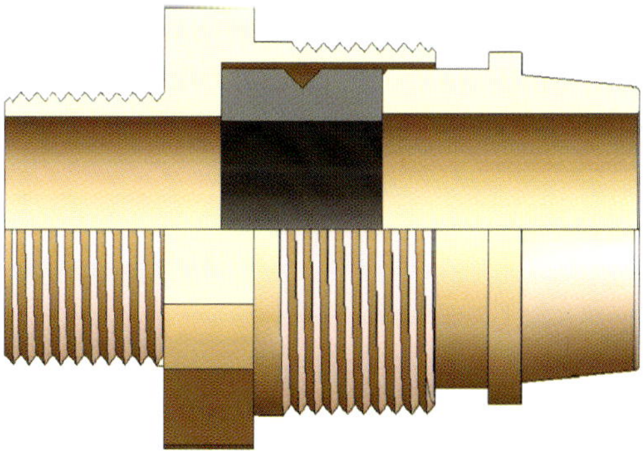

Figure 73
Compression seal

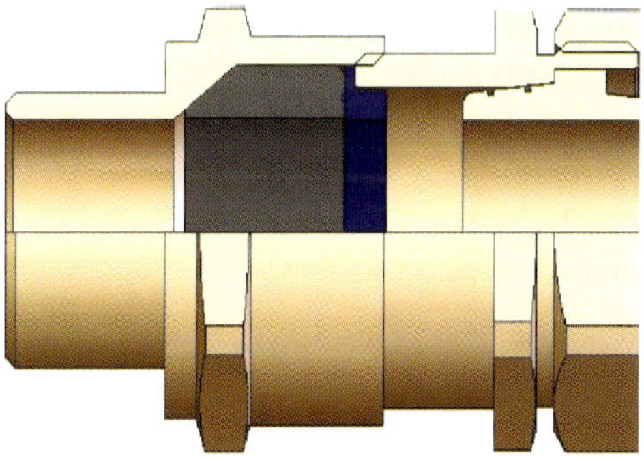

Figure 74
Displacement seal

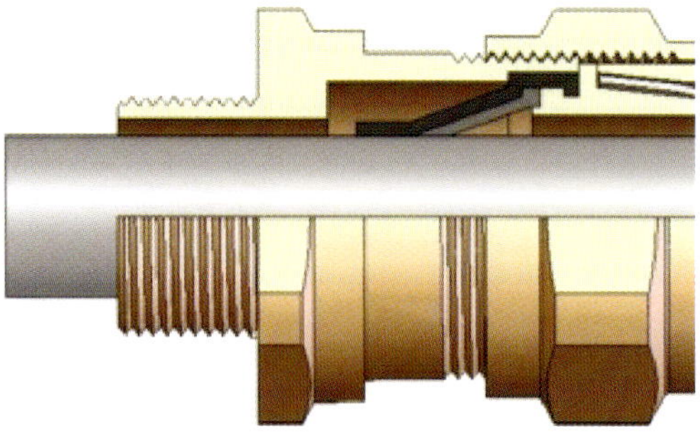

Figure 75
Diaphragm seal

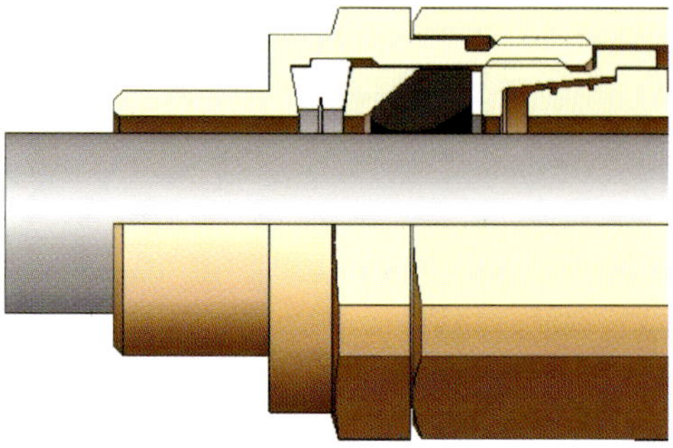

Figure 76
Displacement seal

Compensating displacement

The gland shown in fig. 78 is suitable for unarmoured cable and provides sealing onto the cable sheath and bedding. Once the gas seal is tightened to the necessary force the compensator engages and prevents any further force being transmitted to the inner bedding. With this system the user can employ a face to face tightening of the brass components (Fig. 76).

Compound barrier

This is an additional resin seal applied between the cable cores and is mixed and applied on site. This system is used in potentially explosive environments as shown in fig. 77.

10.4.3. Cable support

The gland shown in fig. 78 is suitable for unarmoured cable and provides sealing onto the cable sheath and bedding. This system also gives good pull out resistance.

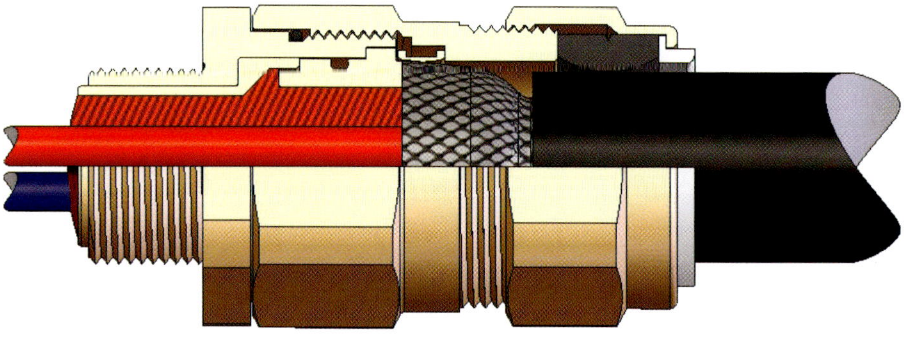

Figure 77
Resin compound barrier

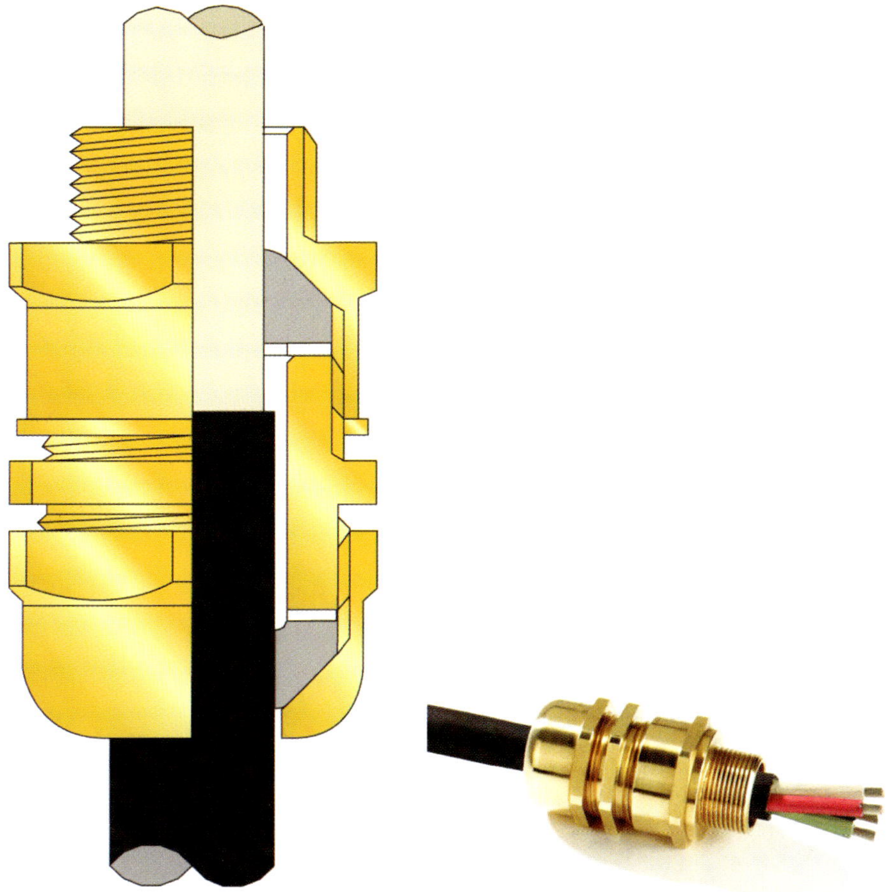

Figure 78
Cable support gland

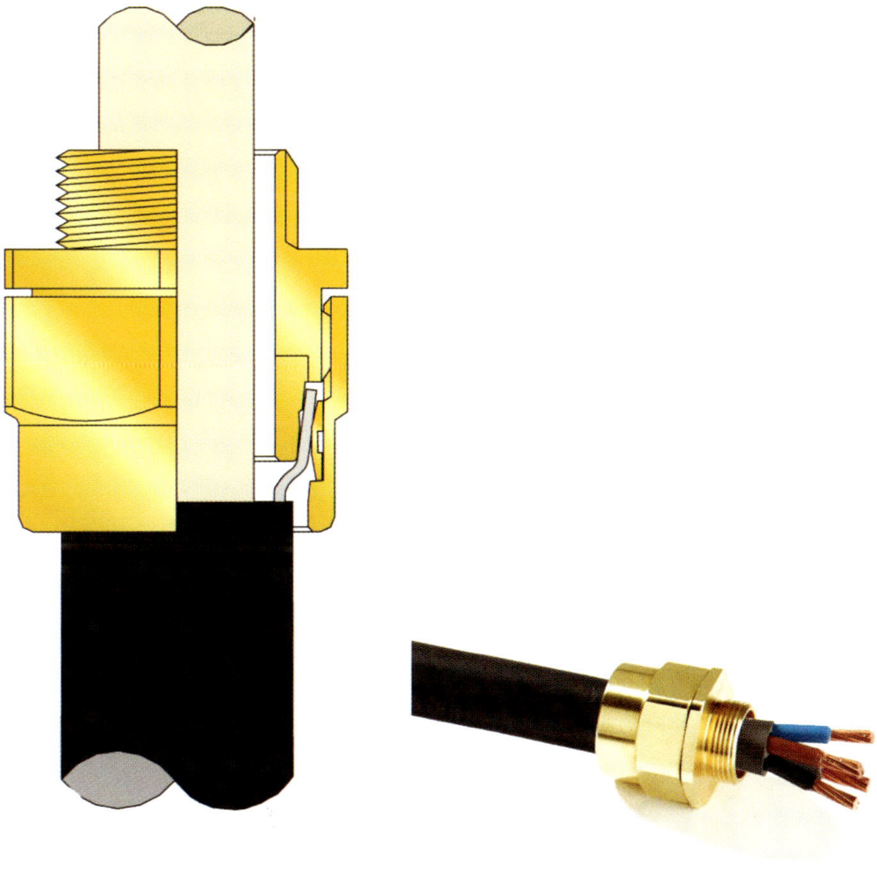

Figure 79
Cable armour
earthing gland

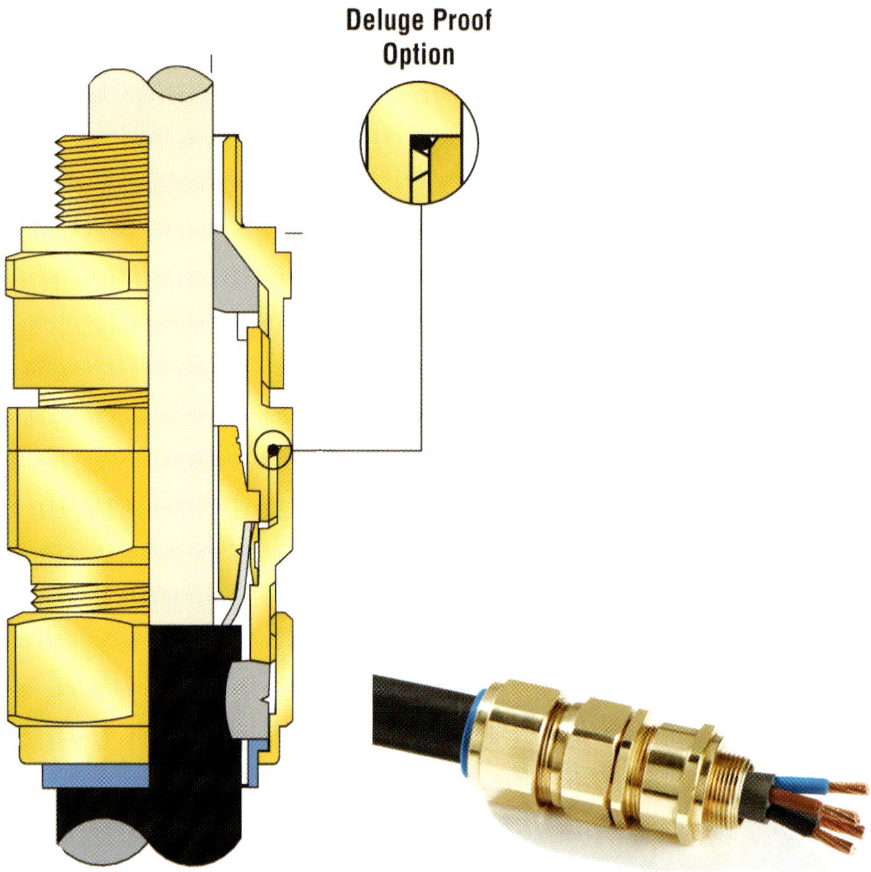

Figure 80
Cable armour
earthing gland
with seals

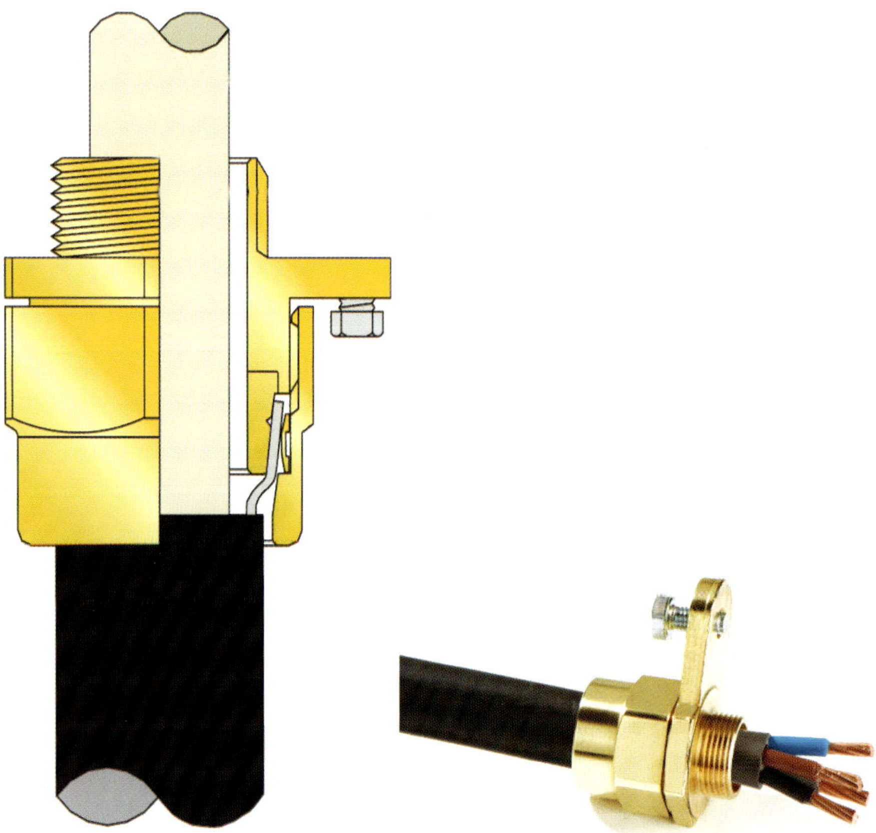

Figure 81
Cable armour
earthing gland
take-off

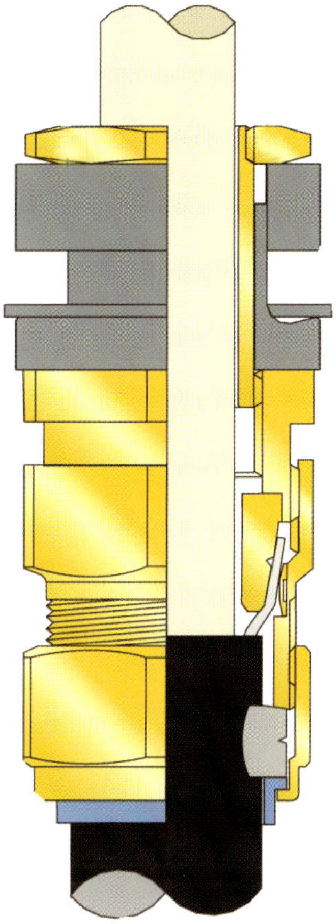

Figure 82
Earth isolation gland

10.4.4. Cable earthing

Figure 79 is an illustration of a simple gland which provides a facility for earthing the cable armours. The cable shown here is LV but the gland can be used on higher voltage ratings. This gland does not provide a seal onto the cable.

The gland shown in fig. 80 will provide seals onto the outer sheath and bedding in addition to armour earthing.

If the short circuit rating of the armours is high it will be necessary to provide an additional earth connection to the gland body to connect a suitable earth lead. Such a gland is shown in fig. 81.

On occasion it is necessary to only earth one end of a cable and this can be achieved using the gland shown in fig. 82.

Chapter 11

Terminations 4
The effects of moisture

11.1. Introduction

The majority of dry terminations will be subjected to moisture during their lifetime whether they are installed in a cable box or are pole mounted. The effect of surface moisture as the result of condensation in a cable box, or rain and drizzle on an outdoor pole mounted termination will be considered in some detail below.

11.2. Formation of a conductive layer on terminations

When an insulating surface becomes moist electric current will flow from the high voltage connection to earth. This current may be of the order of µA up to tens of mA on terminations. This section will consider terminations exposed to the elements to explain what happens.

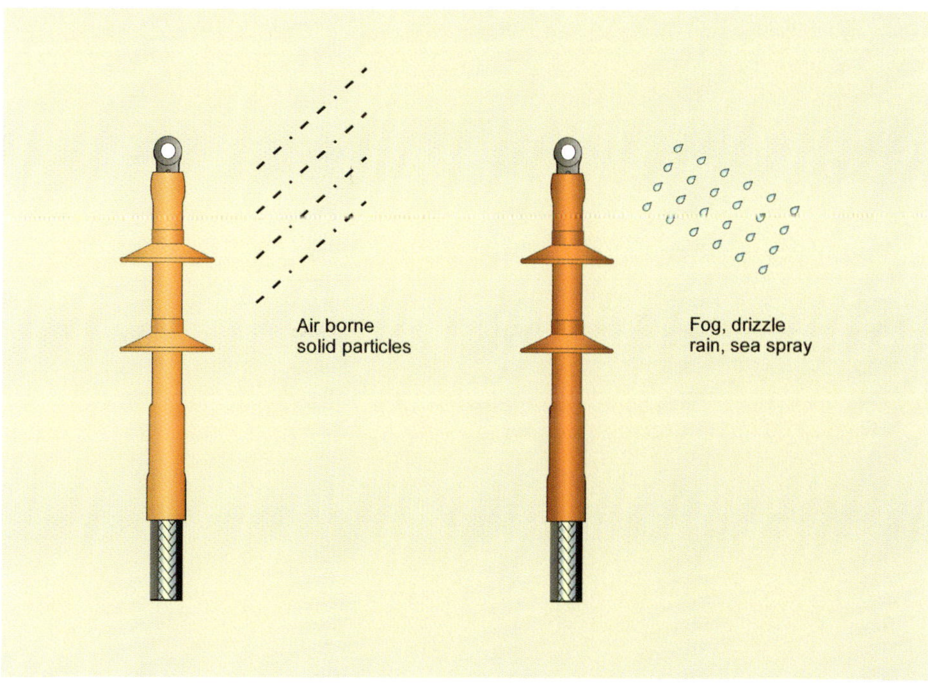

Figure 83
The build up of pollution on a termination

Figure 83 shows how the surface of a single core termination can become covered in a conductive layer. There are basically four types of layer created (usually referred to as pollution) as follows:

11.2.1. 'Clean' moisture

This is moisture, which can result from fog, mist, high humidity and rain. The termination will be situated in a clean, normally rural area, where the surface becomes wet, but the conductivity of the layer is low, as is the associated current flow. Indoor terminations will to a lesser extent be exposed to this type of moisture by way of condensation.

11.2.2. Industrially polluted moisture

Terminations situated in industrial regions will often be covered in a layer of dry solids, which are airborne and build up over a period of time. This layer has little or no effect until it becomes moist when any ionic solids are dissolved to form a conductive layer. This will have a low resistivity. Heavy rain often washes off most of the solids and it is found that the most severe condition is caused when there is fog or light drizzle in the atmosphere, as there is no cleaning effect.

11.2.3. Coastal moisture

Terminations situated near the coast are in the potentially most onerous environment because the air can be laden with salts, which when combined with moisture are highly conductive. It is not unusual for the surface of a termination to have a layer of salt deposited as the moisture carrying it onto the surface evaporates. Successive build up of the salts then produces highly conductive films when they are dissolved in atmospheric moisture.

11.2.4. Condensation

Condensation on indoor terminations can have a far greater effect than moisture deposited on outdoor terminations.

Condensation can only occur if the temperature of the termination surface is lower than the ambient air in the cable box. Cables carrying a significant load continuously will not be affected, however, if the load is very low overnight the termination can cool down to the ambient temperature. As the air warms up in the morning, it will be at a higher temperature than the termination for a period of time and condensation will form. Figure 84 shows how and when this can occur.

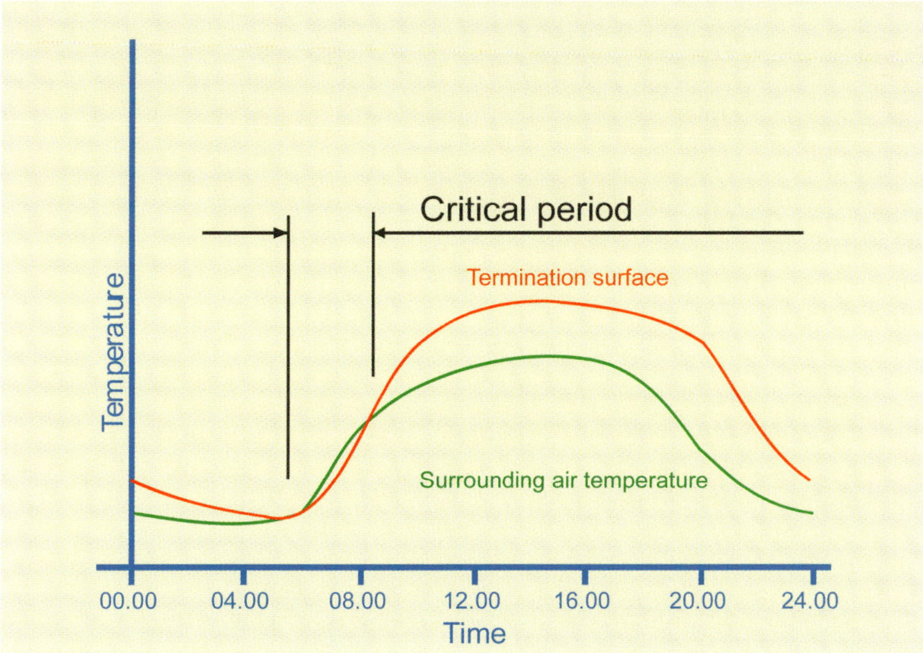

Figure 84
Condition for condensation on indoor terminations

11.3. Effect of a conductive layer

When a conductive layer forms on the surface of a termination, as described above, current will flow as shown in fig. 85a. The effect of this current flow is to heat the moist layer and if the current is high enough it will cause the moisture to evaporate. This evaporation is not uniform, but concentrates in the areas of highest resistance – e.g. at the smallest diameter or a change of profile where there is less pollution build-up.

The evaporation will eventually cause a dry area around the termination since once the drying starts the current flowing has to pass through a smaller area and

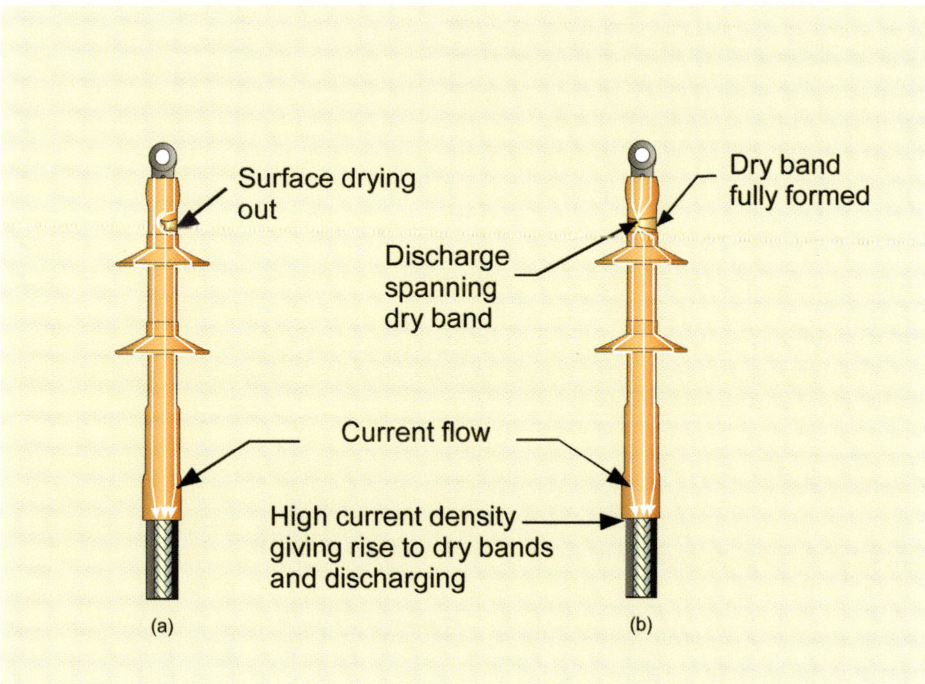

Figure 85
Dry band and discharge formation

subsequently increases the heat generated in that area – this area is usually referred to as a *dry band*. Once the dry band is formed, it can be seen from fig. 85b that the majority of the system voltage is applied across the dry band due to the voltage divider effect. The result of this is that the breakdown strength of air is exceeded and a discharge spans the dry band. This situation is far from stable. The discharge formed may be one of several, which can extinguish or elongate. Generally, the higher the current of the discharge, the more active it is likely to be as it will move around and possibly elongate. Discharges can cause damage in a variety of ways and these will be discussed in detail below.

11.4. The possible effects of discharges on polluted terminations

11.4.1. Surface degradation of the outer material of a termination

When a discharge forms, each end (referred to as a *root*) is in contact with the pollution layer and close to the termination material. The temperature at the root of the discharge can be as high as 1000°C and whilst the area of the root at that temperature is small (a few angstroms) there is sufficient energy to break up a few molecules of the termination outer material. The degradation of the material leads to either tracking or erosion.

11.4.1.1. Surface tracking

Tracking results when discharges cause carbon to be left on an insulating surface as a result of the degradation caused by the heat of the discharge. This carbon deposit will grow to eventually create a continuous conducting path, which will lead to catastrophic failure.

When synthetic materials were originally investigated for possible outdoor insulation applications it was found that many tracked. Early examples of problems included the introduction of epoxy resins to produce insulators exposed to the environment. These insulators had been used in switchgear and thus protected from the environment and had worked extremely well, as they continue to do to this day. However, when subjected to electrical discharges they tracked rapidly. It was realised that the molecular structure of these epoxies was conducive to carbon formation due to the high content of benzene rings, which are highly stable and not readily destroyed by discharge activity. Other constituents within the material would break up and come from the material as a gas. As the breakdown process continued, a carbon path would be created on the surface.

Materials used today for insulators and terminations tend not to track because their molecular structure does not encourage carbon formation when the surface is heated. In addition fillers such as alumina trihydrate are added and when this material is heated it tends to give off steam which has a washing effect on the remaining material. Modern materials also tend to be hydrophobic, which means that they do not wet out easily thus discouraging discharge formation.

Whilst modern materials do not track if exposed to the environment there are situations which can arise whereby tracking will occur. Materials do not track because the by-products of degradation caused by discharges can escape to the atmosphere.

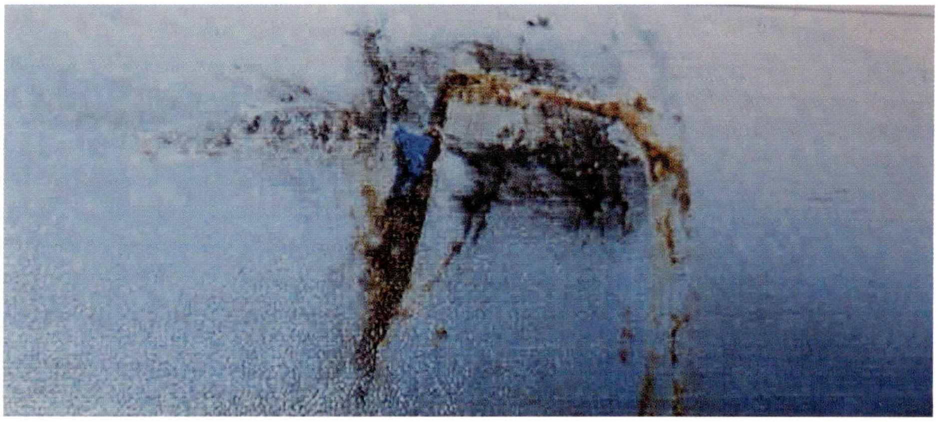

Figure 86
Tracking caused by PVC phase marker

However, if a jointer has added PVC tape as a phase marker, two things happen. Firstly the PVC tape itself will track readily, but it will also degrade. This degradation will allow moisture to penetrate under the tape resulting in the eventual formation of discharges, but as there is no escape for the by-products tracking will occur. An example of this is shown in fig. 86.

Close examination of fig. 86 will show the position of the tape on the termination and the area of black is tracking on the termination surface. This condition was reached within two years of the termination being in service in a desert environment. The use of PVC tapes as phase markers has always been discouraged because PVC tracks very rapidly and it can lead to anaerobic tracking as shown above, but unfortunately the practice continues.

11.4.1.2. Anaerobic tracking

The phenomenon of anaerobic tracking is also an issue when rain sheds are poorly installed on terminations, especially outdoor ones. The sealant applied to the collar of the rain shed must form an effective bond to the termination outer tube otherwise moisture can enter the interface. When this situation exists, dry-bands form under the shed collar and tracking will occur because there is no escape for the by-products when discharging takes place. Figure 87 shows the damage caused to the outer tube of a termination when the rain shed has been removed. This damage occurred in about three years. The termination had not failed at this time, but eventually the outer tube would have been eaten away totally and this would allow moisture to penetrate the interface of the tube and cable insulation. A similar process would then take place along the cable/termination interface and eventually cause total failure.

A further example of this type of problem can be found on indoor terminations. It is usual practice to use an insulating boot to cover the cable box bushing and the termination connection onto the bushing. Provided both ends are sealed to prevent moisture ingress there is no problem, but if the seal is poor and moisture does penetrate, tracking of both the bushing and the inner surface of the boot can occur which will again lead to total failure as shown in fig. 88.

In fig. 88a the bushing shroud had not fully shrunk onto the bushing allowing moisture to penetrate the bushing/shroud interface. The tracking is clearly seen on

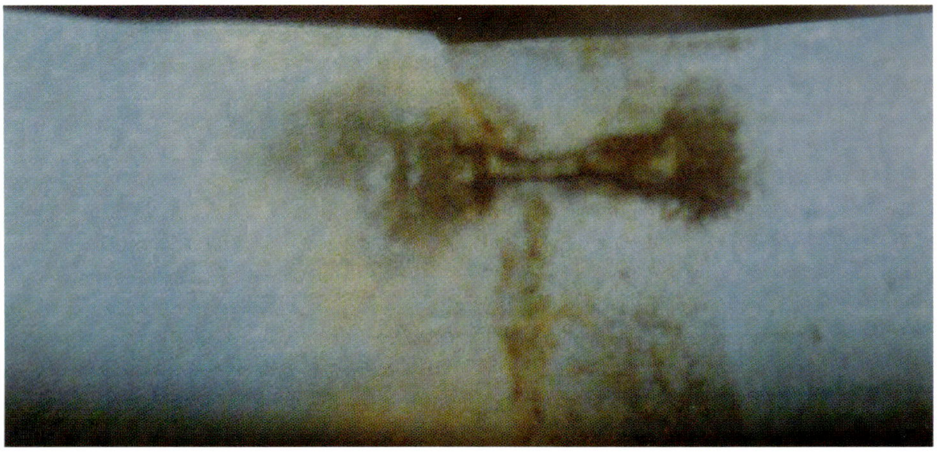

Figure 87
Tracking under a rain shed

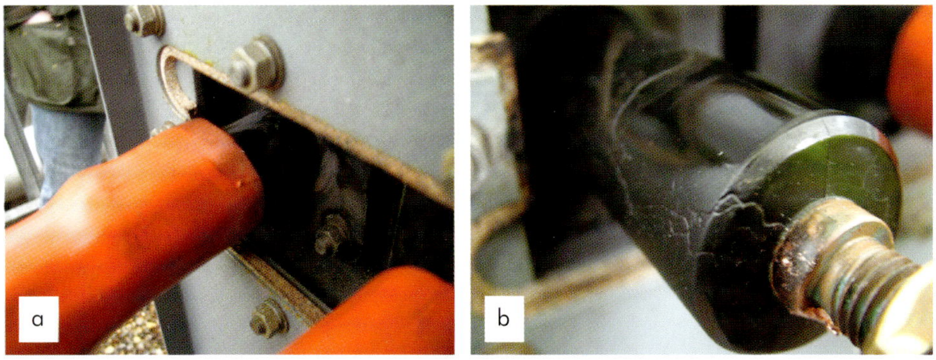

Figure 88 (a) and (b)
Example of tracking caused by poor seal onto a bushing

the bushing in photograph (b). Note that the bushing had been cleaned of the debris of the flashover prior to the photograph being taken.

11.4.1.3. Surface erosion

Erosion of a material is caused when surface discharges destroy the molecular structure, but do no leave a conductive (carbon) residue. Modern materials such as silicones and EVA used in dry terminations will erode at such a low rate that they are likely to continue to operate satisfactorily for well over 50 years. At the time of writing there are terminations still in service that were installed almost 40 years ago and remain in perfectly good condition.

12.4.1.4. Surface flashover

The likelihood of surface flashover occurring on an outdoor termination is highly improbable, but will be mentioned here for completeness. Insulators used in transmission and distribution networks are a source of flashover, because they are made as small as possible to reduce the size of support structures and hence they have relatively short creepage distances. A termination, on the other hand, has no such size restrictions and hence the creepage distance is always much longer than on the remainder of the system.

Surface flashover, if not caused by a system over voltage (e.g. lightning strike), can occur if the surface becomes polluted. As discharges form and the conductivity of the pollution layer being high, they can elongate by running along the surface of pollution and join up with other discharges in series until the live and earth is spanned. In the case of terminations, the creepage length (and hence surface resistance) between live and earth is such that the discharge currents are restricted to a few mA and are not able to elongate significantly.

11.5. Sealing of terminations

It is essential that water is not allowed to enter a termination and this requires that any interface should be effectively sealed. Seals are usually provided on heat shrinkable terminations by means of hot melt sealant applied to the inside of the component parts. Tapes can also be used and whilst they take a little longer to apply, do not require heat to activate the seal. Any components recovered sufficiently to be in contact with the tape will provide a seal. In the case of hot melt sealant sufficient heat must be applied to produce a satisfactory seal. Cold applied terminations will often use a tape sealant and rely on the hoop stresses in the components to maintain an effective seal. In the case of push–on terminations lubricant such as silicone grease will not only ease installation, but also provide a seal.

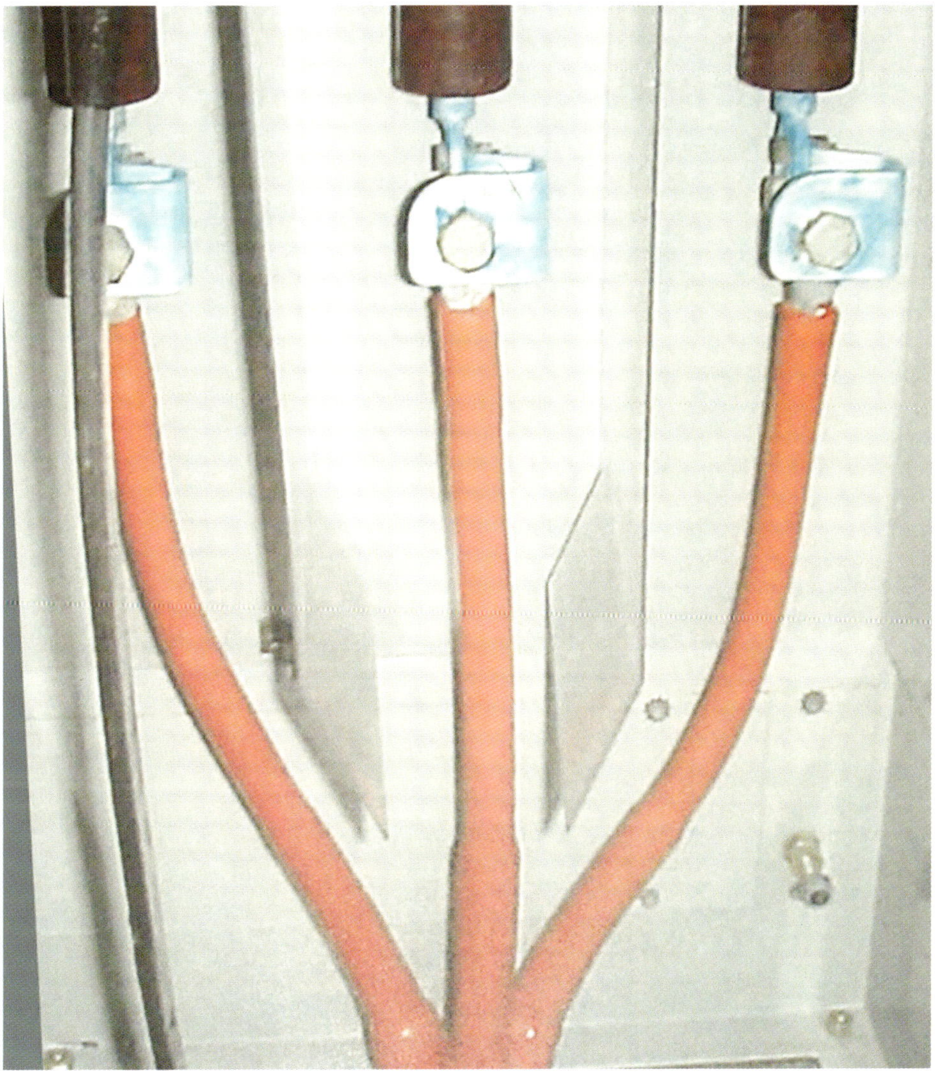

Figure 89
Example of discolouration caused by discharge

Figure 90
Example of dry band formation

Failure to seal the termination effectively will allow moisture to enter, and once it has spanned live and earth will create anaerobic tracking explained in the previous section.

11.6. Reducing the effects of surface discharges

It has already been stated that outdoor terminations are not affected greatly, provided they have been correctly installed. However, with indoor terminations this can be a major problem. Before ultimate breakdown there are tell tale signs. In fig. 89 it is the copper turning green. In other instances the termination will be discoloured due to dry band formation (fig. 90). The light coloured areas in fig. 90 are deposits of alumina trihydrate released when the base polymer has eroded due to discharging.

The best way to prevent a problem due to condensation is to use a screened separable connector since the surface is at earth potential. This would not be possible in the case shown in fig. 89 therefore it is recommended that a silicon rubber cold shrink system be used as the surface does not readily wet out. Any condensation is in the form of discreet droplets therefore a continuous current path does not exist.

Chapter 12

Joints – An overview

12.1. Introduction

Many of the issues discussed so far relating to terminations are equally relevant to joints. Essentially, the function of a joint is to electrically connect the conductors of two cables and then replace all the components removed from the cable, in order to effect a safe and reliable connection. The various functions of a joint are listed in the table below:

Function	How achieved
1. To join conductors	Using connectors compatible with the conductor metal. The two main types are crimped and mechanical. They can be blocked if one cable is paper insulated.
2. To exclude air between high voltage and adjacent insulation	Using systems similar to those for a termination and in addition a Faraday Cage over the connector can be used.
3. To provide stress control	As with a termination, also can use a Faraday Cage over connector to help field distribution.
4. To replace primary insulation	Can be separate tubes or an integral part of a moulded component.
5. To replace earth screen	A conductive material will usually be extruded or moulded onto the insulation.

Function	How achieved
6. To replace metal earthing system	Can be performed using a copper stocking, copper braid or metal canister depending upon short circuit requirements.
7. To allow cross bonding of single core cables	An earth lead is taken from the joint.
8. To seal against moisture ingress	Seals can be of the same type mentioned in the section on termination sealing.
9. To provide mechanical strength	There will be some component within the joint which provides this. It may be in the form of a metal canister or several hybrid joints today will incorporate resin.

12.2. Connecting the conductors

Crimp connectors have for many years been the most popular, but mechanical designs employing bolt heads, which shear off at a certain torque, are becoming universally acceptable. An advantage of the latter type is that no expensive crimping tool is required and there is no distortion of the profile. Range taking is also a feature not present in basic crimp connectors, which are conductor size specific. In addition there are a range of connectors which are in two halves, which makes installation much easier, especially on three core cable. However, the cross section of the mechanical connector is not always symmetrical and is likely to be of a larger diameter than the crimped variety. The significance of this will be discussed at a later stage.

The subject of connectors can become quite involved and it is not within the remit of this book to proceed beyond basic descriptions. For medium voltage a good connector will have a smooth profile which will have slightly tapered ends.

12.2.1. Crimped connectors

Connectors used on MV cables are shaped to provide a profile which does not lead to excessively high stresses. A typical connector for copper conductors is shown in fig. 91. The aluminium equivalent will be longer because of its inferior conductivity.

Figure 91
Copper connector for MV joints

There are two basic dies used for crimping these connectors, namely indent and hexagonal. The former places indents along the connector and the depressions made are then filled with a putty type material. The latter will create hexagonal pressure around the connector. In this case it is likely that sharp edges are formed and these must be filed down to give a smooth profile. In both methods, cold welding

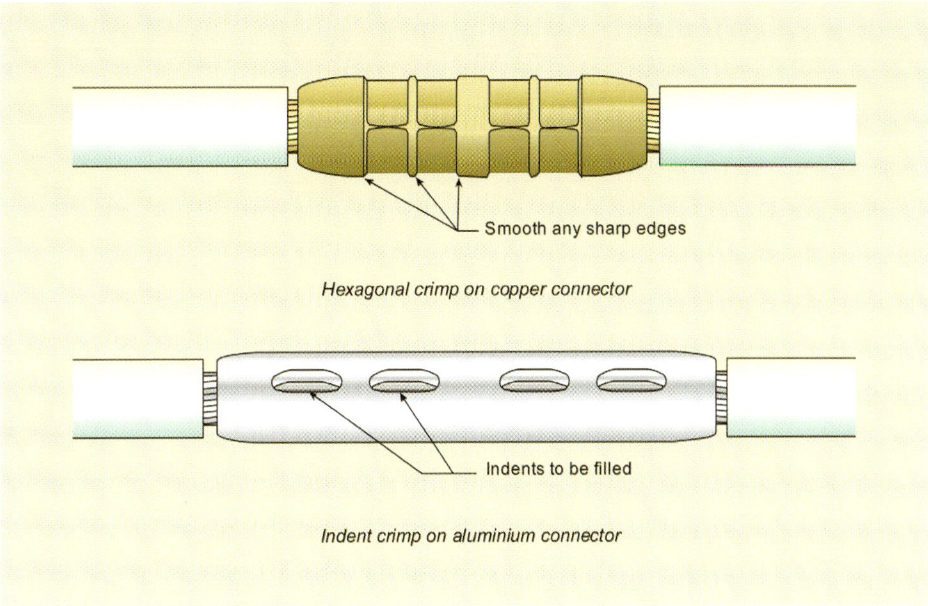

Figure 92
Two types of crimp in common use

takes place between the connector and the cable conductor to give good electrical contact.

It will be found that copper connectors are somewhat shorter than their aluminium counterparts due to the superior conductivity of the copper. Typical crimp connectors are shown in fig. 92.

12.2.2. Mechanical connectors

Mechanical connectors have the following advantages over the crimp system:

- They are range taking.
- They do not require expensive tooling to install.
- The jointer cannot use the wrong die; all that is required is that the bolt heads shear off.
- The 'Split' types are easier to install especially on three core cable because the cores to not have to be bent to put the conductors into the connector.
- 'Y' or branch joints can easily be accommodated.

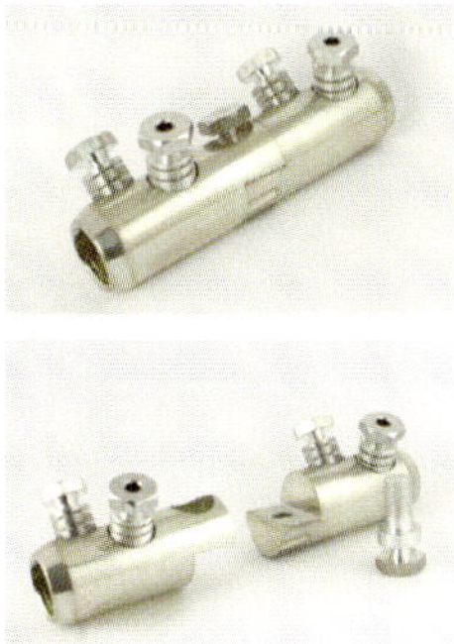

Figure 93
A typical split type mechanical connector

A typical split connector is shown in fig. 93. Note that each end is blocked, which, when jointing a paper cable to a polymeric, prevents oil migration from the paper cable.

The three disadvantages of this type of connector are that:

- It may be considered expensive, especially if the end user is fully tooled up for crimping.

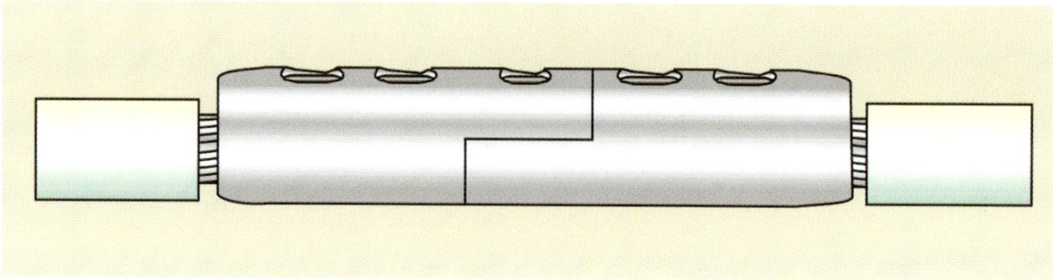

Figure 94
Installed non symmetrical mechanical connector

- It has a larger profile than the equivalent crimp connector, therefore all systems may not easily be installed, especially push-on types.

- It is not always symmetrical in cross section, which can make treatment over the connector area more critical, see fig. 94.

Methods of overcoming this asymmetry will be discussed in the next section.

12.3. The elimination of air from critical areas

It will be seen from the above diagrams that the area surrounding the connector requires attention to eliminate the effect of air in what is a potentially highly stressed region. There are two basic methods of doing this as follows:

12.3.1. Stress relieving tape

Stress relieving tapes can be wrapped around the connector to exclude all potential voids and build up a smooth profile over which subsequent layers of the joint can be installed. This method is effective, but does require some skill on the part of the installer. A connector that has been taped is shown in fig. 95. It will be evident that this is more easily applied on a crimped connector than a mechanical type because less tape is required to build a smooth profile. The final profile must be such that any subsequent components applied maintain total contact with this tape and does not allow air to be trapped. This technique is most commonly used with heat shrinkable joints.

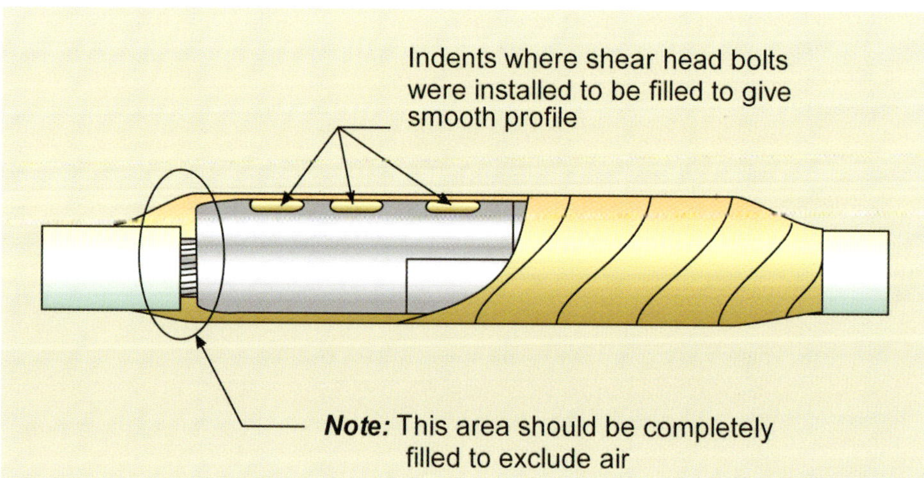

Figure 95
Connector covered with stress relieving tape to give smooth profile

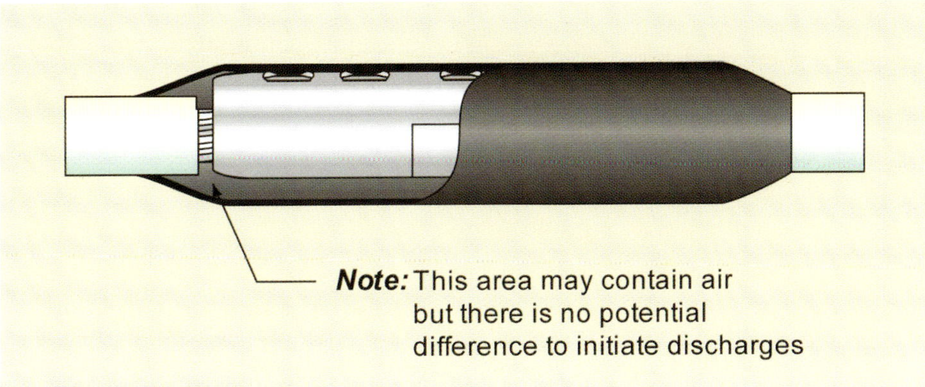

Figure 96
A Faraday cage positioned over a mechanical connector

12.3.2. Faraday cage

This method does not remove air, but simply ensures that it is in a region which is not electrically stressed. The Faraday Cage will usually be in the form of a conductive rubber often as an integral part of the joint. Part of the cage must be in contact with the connector in order that the whole is at the potential of the conductor. Technically, this is a better system and less skill is required to install the joint. The Faraday cage will ensure that a smooth profile is obtained and in direct contact with the insulating layer. The total effect is an optimum electric field distribution and entrapped air in critical areas is not under stress.

It will be noted in fig. 96, which shows a typical Faraday cage, that there will be air between the ends of the connector and the cable insulation, but the potential across the air is zero due to the effect of the cage and discharging will not occur.

The Faraday cage shown above can take various forms depending upon the jointing system. It will usually be an integral part of the joint in order to ensure there are no voids between the cage and the insulation. In order to achieve contact with the connector there will be raised portions on the inside to ensure good contact is maintained on the smallest connector in a given application. Provided contact is made with one side of the connector there is no problem with air between it and the Faraday cage since both are at the same potential. Similarly, it is not important to fill the gap between the cable insulation and connector as there will be no potential difference in this region.

12.4. Stress control

This can be achieved using a geometric system or with the use of high permittivity material positioned across the joint from one earth screen to the other. The following will refer to polymeric insulated screened cables.

Geometric type – a) Push on

This method is used in multi layer moulded joints where a Faraday cage is included in the centre of the joint. The outer conductive layer is moulded over the insulating layer and provides a geometric profile which gives stress control at the screen cuts as shown in fig. 97. This is an early design for a push on joint and this system works very well since there is no need to treat the area over the connector and careful

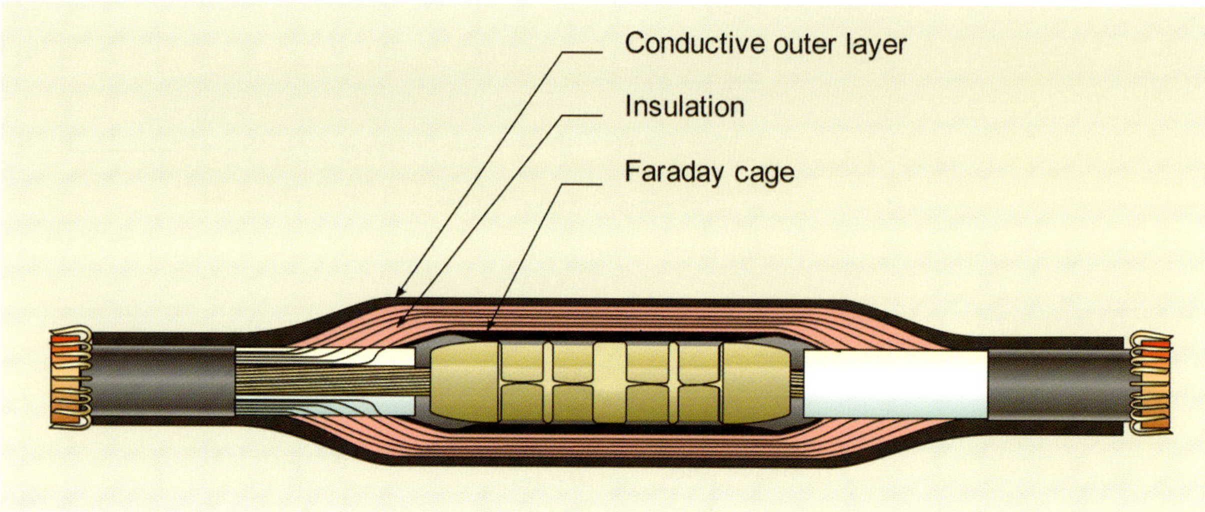

Figure 97
Cross-section of a push-on joint with electric field shown

design provides a reasonably uniform electric field. There are some logistical drawbacks to this system as follows:

- This type of joint is less able to cope with range taking than shrinkable systems.
- If it is to be used for both aluminium and copper cables the Faraday cage must be long enough to cover the longest aluminium connector, thus making the joint longer than it need be for copper cable.
- Care must be taken when positioning the joint to ensure that it fits centrally over the connector.
- It does not readily lend itself to three core cables because of its bulky cross section and in addition it is necessary to install other, non push-on, components to seal the cable against moisture ingress. This type of joint is most widely used in the United States and other countries adopting their system on single core URD cable networks.

It will be noticed that the electric field lines are contained inside the outer earth layer of the joint and the proximity of the earth layer will affect the electric stress in the joint. The more distant the earth layer is from the cable the lower the stress in the joint. Clearly the thickness of insulation dictates the stress in the joint and the

Figure 98
Cross-section of a cold shrink joint

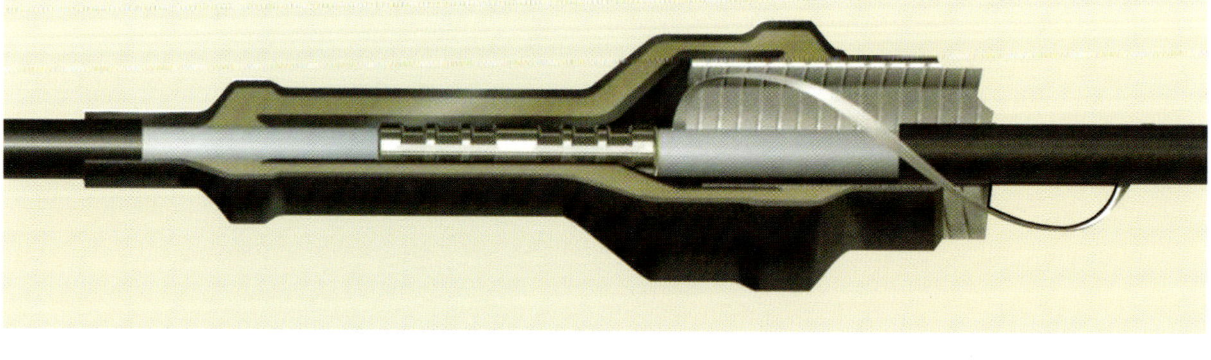

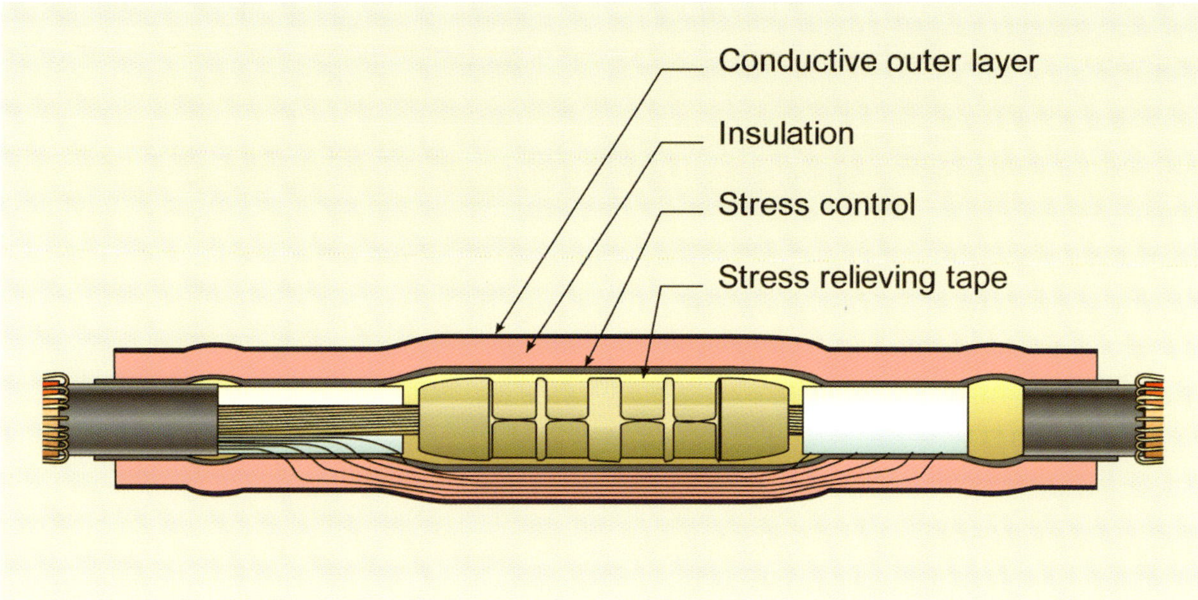

Figure 99
Cross section of joint using high permitivity stress control

thickness is usually a little greater than that of the cable insulation. Making the insulation any thicker does not provide any advantage, but would increase the cost.

Geometric type – b) Cold shrink

The most popular system employing a Faraday cage is cold shrink. The joint is on a holdout which is removed when installing to place the joint over the connector and cable on either side of the joint. See fig. 98.

12.4.3. High permitivity

This method employs the same technology as for a termination and will consist of a high permittivity tube. This tube can either be supplied as a single component or it can be an integral part of the joint having been moulded to the insulation layer or co-extruded onto it, depending upon the manufacturing process.

Figure 99 shows a joint employing this system.

In the above drawing the electric field is only shown on the lower half of the joint for clarity. This type of joint will normally be of the shrinkable type and in fig. 99 a separate stress control sleeve, which is slightly longer than the insulating tube, is shown. The outer conducting layer will normally be co-extruded or moulded onto the insulating layer. It is possible to employ a triple extrusion whereby the stress control layer is an integral part of the tube.

12.5. Thickness of insulation

All shrinkable tubes, when expanded, exhibit a reduced wall thickness. The manufactured thickness is only achieved if the tube is allowed to fully recover. Any tube used on a joint will not fully recover in order to exclude all air and provide a seal at each end of the joint. As a result, when designing a joint, the recovered insulation thickness must be considered and must be of adequate thickness on the largest cable in the range being covered. As a rule of thumb, the minimum recovered thickness of insulation over the connector will be about 20% greater than the insulation thickness of the cable. This is a safety margin to allow for less than perfect installation, since the electric strength of the insulating tube should be at least as high as the cable insulation.

The recovered insulation thickness on cold applied products will be consistent around the joint, but with heat shrinkable systems the installer must ensure uniform heating around the whole tube. Excessive heat on one side will result in the insulation thickness varying quite significantly. In the worst case the thickest section could be almost twice the thinnest, with the result that the latter would be under a much higher and possibly damaging stress.

Figure 99 shows a single insulating tube with a conductive outer layer. It is necessary sometimes, at higher voltages, to have a second insulating layer to build up the required thickness. This tube is installed on top of the stress control tube with the insulating/conducting tube applied on top.

12.6. Sealing the joint

The joint must be sealed against moisture ingress otherwise failure is inevitable. It is accepted that on three core polymeric cables there is the possibility of water entering the joint area as a result of sheath damage to the cable away from the joint. This is not in itself a major problem provided the joint is sealed but if water enters the joint problems will occur. The following sections only consider the seal around the joint area, but in practice, most of these joints will have an outer sleeve to provide corrosion resistance and some mechanical protection. These sleeves will be sealed onto the cable sheath with either a hot melt adhesive, in the case of heat shrink tubing being used, or a self amalgamating tape for cold shrink.

12.6.1. The one piece joint

This type of joint will have been produced in such a way that there is only one interface to be sealed as shown in fig. 100.

There are two types of failure possible in this arrangement. If the seal is very poor and allows unimpeded flow of water (see the left side of the drawing), it will quickly short out the conductor and earth and an interfacial breakdown will occur.

Figure 100
Effects of moisture ingress in a one piece joint

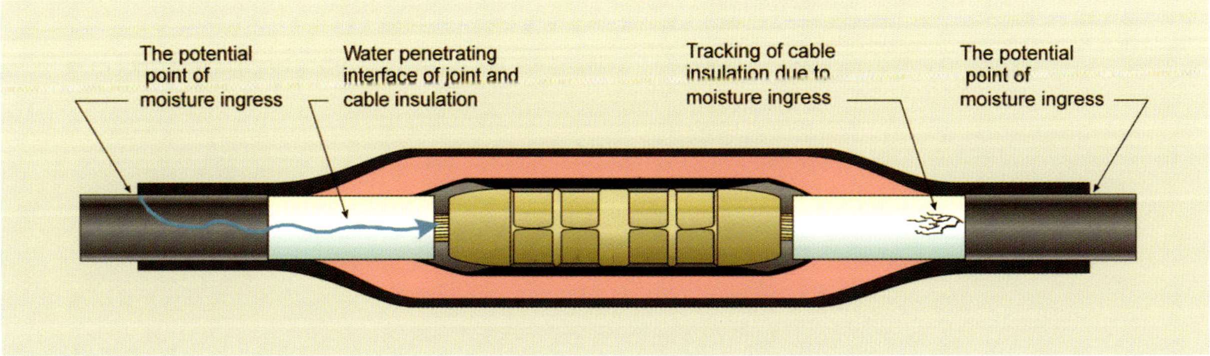

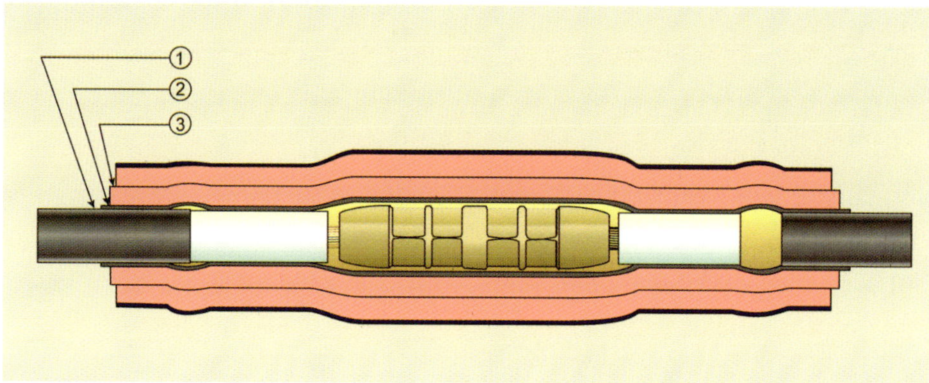

Figure 101
Areas of a joint which must be sealed

When the seal is only allowing limited moisture to enter the joint discharges will occur along the interface of the joint and cable insulation and this in turn will lead to anaerobic tracking. Eventually the tracking will proceed until the space between it and the conductor will break down and then complete failure will occur.

Sealing of this joint is usually reliant on the hoop stresses imparted by the joint onto the cable. Silicone grease is normally used to help install the joint and this also helps to provide the seal and prevent moisture ingress. Generally this joint does not have a problem with sealing and the most likely causes of the seal failing are either the wrong size joint has been installed, or foreign bodies have been trapped during installation.

12.6.2. The multi layered joint

This type of joint has potentially more chance of suffering from moisture ingress simply because there are more interfaces. Figure 101 shows a multi-layer joint and the various types of failure that can occur due to poor seals. Two insulating layers are shown simply to illustrate all possible failure modes.

The joint components will be heat shrink or cold shrink and rely to a certain extent on the hoop stresses to give the seal. This is particularly true with the cold shrink system. Heat shrink systems apply considerable hoop stresses during shrinking, but once cooled no longer apply pressure. It is important therefore to fully shrink the components in place. The most modern heat shrink joints employ an elastomeric insulation layer with a heat shrink outer layer and with this type of product there will be hoop stresses after shrinking is complete.

Figure 101 – Areas of a joint which must be sealed

The three points on the above joint must not allow moisture ingress as follows:

1. The cable/stress control interface.

If moisture penetrates this interface, breakdown will occur in the same way as explained in the one piece joint in fig. 100. The stress relieving tape applied at the screen cut will normally ensure a good moisture block and therefore a problem here is unlikely.

2. The stress control/insulating layer interface.

Moisture permeating along this interface will lead to breakdown through a mixture of tracking and puncture of the stress control layer.

3. The insulation/insulation interface.

The normal breakdown process in this interface is for tracking to occur, which proceeds along the interface towards the centre of the joint. At some point this will cause a volumetric breakdown of the insulating layer through to the connector. Many modern joints overcome this problem by using a single, thicker insulating layer.

The breakdown of joints due to moisture ingress is not too common, but does occur with badly installed joints. If the design is inherently weak, failure will occur after a few cycles of the type testing program when the joint is load cycled under water. This test would thus disqualify it from being considered for use in the field. As a precaution in the field, a tape sealant could be applied between layers provided it is outside the stressed areas of the joint but this is not usually considered necessary.

A further method is to smear silicone grease between layers and this can help some heat shrinkable joint designs where different longitudinal shrinkage of components can, in very extreme cases, cause rippling of the tubing and void formation. The use of grease allows components to slide freely as they shrink. However, a product that has been correctly installed will generally work satisfactorily without additional sealing.

12.7. Earthing

The earthing system should be such that it does not significantly alter the earth capabilities of the cables that have been jointed. The basic methods of making an earth connection in a joint are similar to those for a termination. The main difference is that in this instance an earth connection has to be made from one cable to the other. Paper cables will not be considered at this time but will be covered in a following section on hybrid joints.

This section will consider some earthing methods available for joints and will examine the requirements for single core cables. It is important to address this because it is usually the responsibility of the accessory manufacturer to provide an

Figure 102
Copper mesh shield being applied across a joint

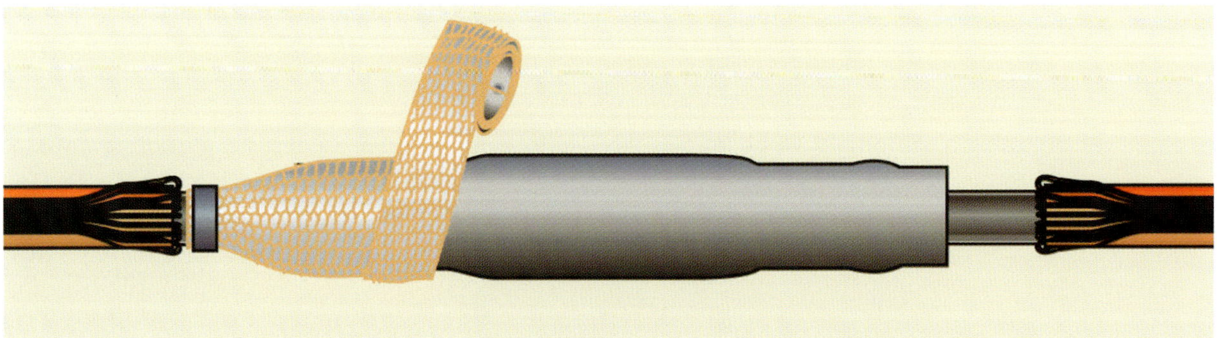

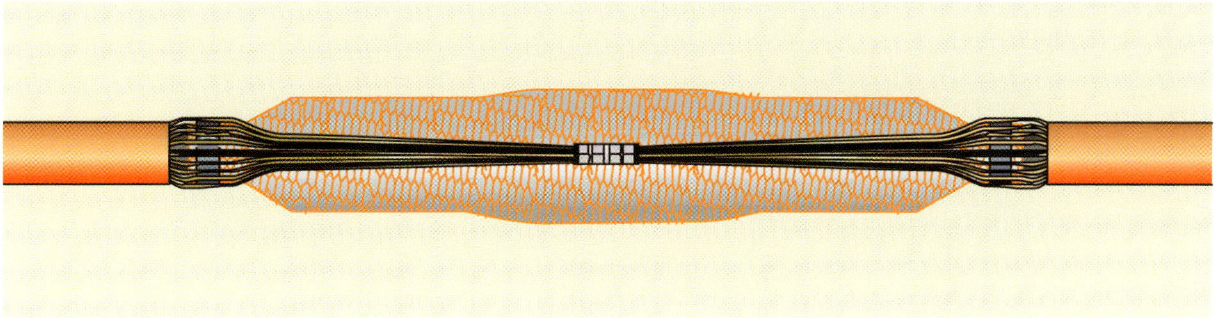

Figure 103
Complete earth connection using copper mesh and cable screen wires

adequate earthing system. The end user and the accessory supplier should ensure that they are both aware of the worst case scenario relating to circulating currents in the cable sheath and armours and the short circuit current.

12.7.1. Single core

Reference to the earth layer on joints has been mentioned in previous sections. This earth layer in effect is performing the same function as the conductive screen on the cable insulation. The push-on joint conductive outer layer in fig. 97 will be in contact with the cable extruded earth screen, but the type shown in fig. 99 does not show a connection between the cable earth and the conducting layer of the joint. Such a connection must be made to ensure that the conductive layer is at earth potential. There are a variety of methods of achieving a good connection depending upon the cable construction, but the commonest method is to wrap a tinned copper mesh across the joint. The mesh is in contact with the outer conducting layer of the joint and is connected to the cable earth at each end. Figure 102 shows such a method.

The mesh shown above is in contact with the extruded cable screen and additional contact will be made by way of the screen wires as they are folded back over the joint (see fig. 103).

The above earthing arrangement is one of the more simple types for a single core cable and to complete the joint it is simply necessary to fit an outer cover to seal against moisture ingress. The earthing of armoured cables requires that an earth strap spans the joint and is connected to the armours of both cables.

12.7.2. Three core joint earth

There are as many variations of earth connections as there are cable constructions, but they all perform the same task which is to continue the earth connection across the joint. Figure 104 illustrates one common type where the cables are steel wire armoured.

The main earth current carrying conductor across the joint is the copper braid and must be of sufficient cross section to ensure that it can carry the system short circuit current. The metal canister, which is principally to provide mechanical strength, will also carry some current.

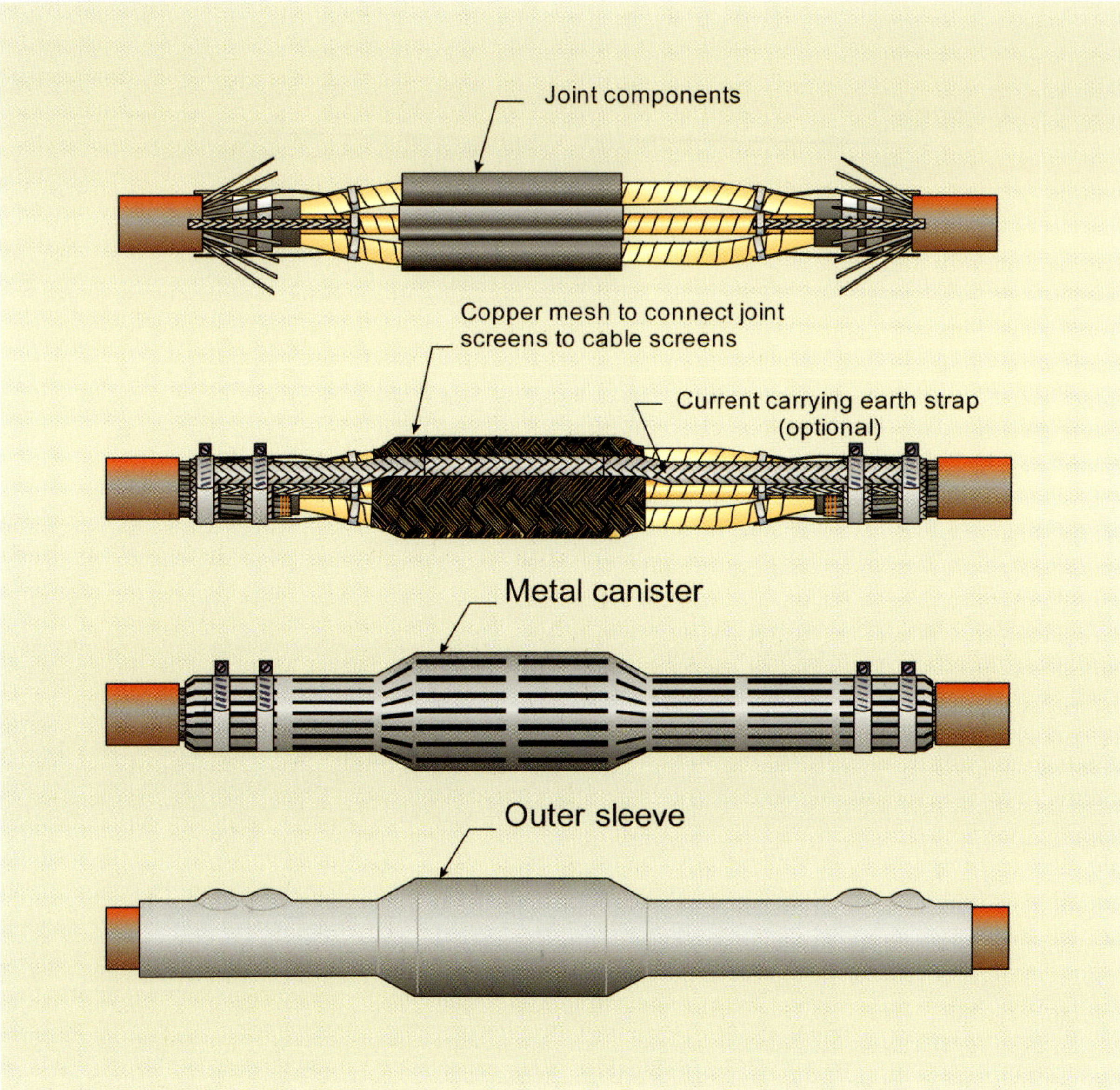

Figure 104
Earthing for three core steel wire armoured cables

12.8. Mechanical strength

Heat shrink and cold shrink joints used on single core cables will rely on the shrinkable components to provide sufficient mechanical strength. Single core joints are usually quite short but there is loss of mechanical strength when compared to the cable. It is important therefore that the joints are supported.

Three core cable joints do not lend themselves readily to one specific technology, be it heat shrink or cold shrink, for various reasons and hybrids have been developed which are usually resin filled. These joints are mechanically strong and abrasion resistant.

12.9. Hybrid joints

As mentioned above, when dealing with three core joints the best designs use more than one technology. Early versions of heat shrink and cold shrink joints were theoretically, when correctly installed, perfectly fit for purpose. However, there were several drawbacks:

1. They required a high degree of skill to install correctly.
2. They were very long.
3. Mechanically they were relatively weak if direct buried.
4. They were not suitable for branch joints.
5. Sealing against moisture ingress could be difficult.

The hybrid joints were developed to provide an improved design to eliminate the above problems and were encapsulated in polyurethane resin. It is also possible to connect various cable combinations.

The two types most commonly found today are shown in the following sections.

12.9.1. Moulded Faraday cage over connector

The joint shown in fig. 105 connects three core paper belted cable to three single core polymeric cables. Any combination of cable types can be jointed with this system.

It will be noted that the mechanical connector used in this design is the low voltage type because the red moulding over the connector contains a Faraday cage – see fig. 106.

The paper side of the joint has clear heat shrink tubing over the cores and at the breakout a heat shrink collar is applied. The ends of the earth screens are taped to provide some stress control. After the metal canister and moulded outer case is fitted around the joint it is filled with polyurethane resin. The resin provides electrical insulation, moisture seal and mechanical strength. This particular design is suitable for voltages up to 15 kV.

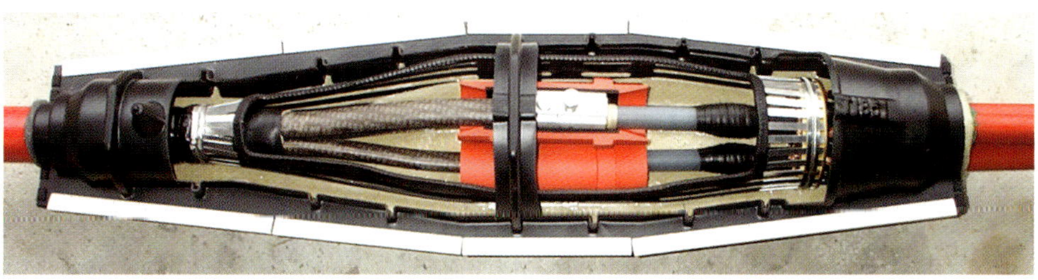

Figure 105
Cross section of hybrid joint

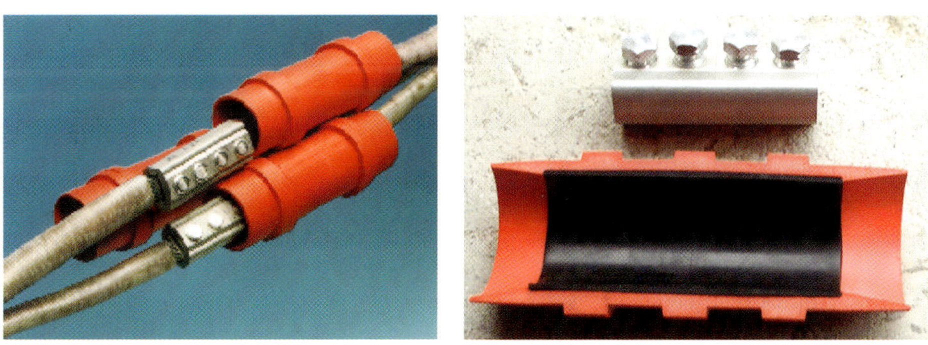

Figure 106
Connector area of joint and moulded connector cover

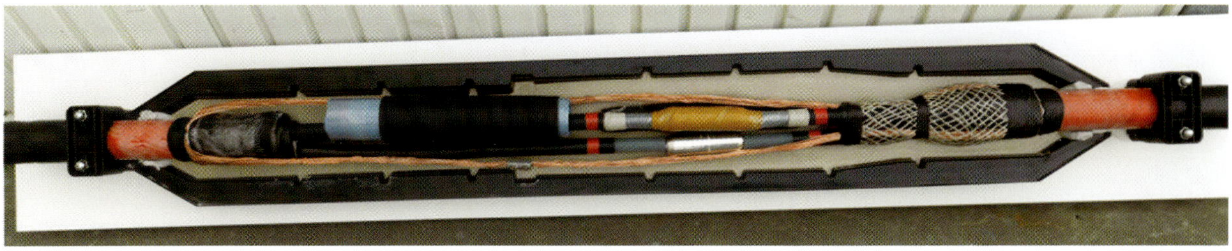

12.9.2. Shrink tubes and resin combination

Figure 107
Hybrid joint incorporating cold shrink tubes over the connectors

This design is versatile and can be used on system voltages up to 36 kV. Figure 107 shows a typical three core joint. The area over the connector is covered with a stress control patch and a dual wall cold shrink tubing, consisting of an inner insulating layer and outer conductive layer, is shrunk over the connector. With this system the connector must have a smooth and shaped profile. The conductive layer is attached to earth thus there is no electrical stress between the cores. The earth across this joint is a combination of copper wires and a copper stocking. The joint is then encapsulated in a resin.

Similar joints are produced using a heat shrink dual wall tube.

The above system can also be used to produce a trifurcating joint as shown in fig. 108.

Note how the steel wire armours of the three core joint are connected to the three single core copper earth wires of the single core cables.

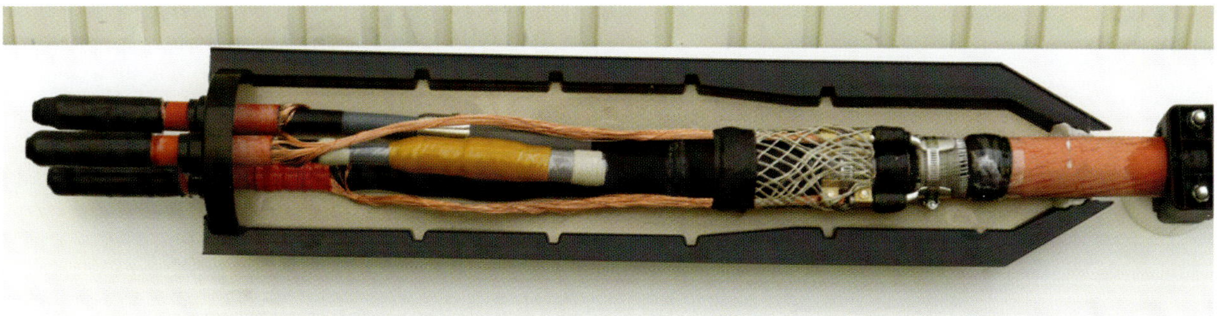

Figure 108
Hybrid trifurcating joint

Chapter 13

Earthing of single core cables

13.1. Induced voltages and currents in single core cables

When single core cables are earthed at one end only an induced voltage will appear at the other end. Earthing at both ends will result in circulating currents. It is necessary to consider the following before deciding on the best course of action.

When both ends are earthed is the circulating current excessive?

When only one end is earthed will the induced voltage at the other end be a problem?

13.1.1. Induced voltage

The induced voltage on a single core cable will be a function of the following:

- The current in the cable conductor.
- The frequency of the current (usually 50 Hz or 60 Hz).
- The length of the cable.
- The proximity of the other two cables (assuming a three phase system).
- The mutual inductance per phase.

The voltage, V, induced between the conductor and the sheath is given by:

$$V = I\omega M \, volts/m$$

Where,

I = the current in the conductor
ω = 2πf (314 for 50 Hz system)
S = separation of the cables
d = cable diameter

$$M = 0.2\ln\left[\frac{2S}{d}\right] \mu H/m$$

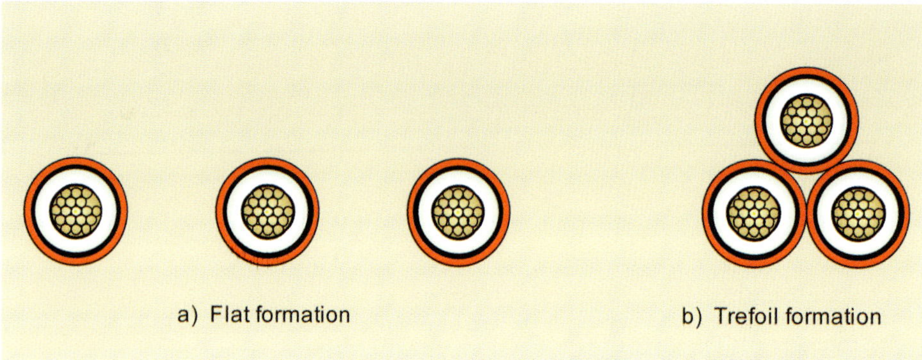

Figure 109
Single core cable formations

The total voltage induced will be the sum of the voltages induced by each of the three phase currents. The closer the three cables are to each other, the lower the induced voltage because it is the vector sum of the three phases. The worst case is where the cables are laid in a flat formation, fig. 109a, and are widely separated. In this case the centre cable will have the least induced voltage.

The arrangement to give the lowest induced voltage will be when the cables are in trefoil, fig. 109b, and this will be a minimum when the cables are touching.

Triplex cable, which consists of three single core cables with a twist, the same as for a three core cable, will exhibit a much reduced induced voltage because the electric field produced by the three cores cancel each other out.

As a rule of thumb for a touching trefoil arrangement, for every 100 amps of load current, 4 volts per kilometre is induced.

13.1.2. Induced current

When single core cable is earthed at both ends current will flow in the sheath as a result of the induced voltage. This current is calculated as follows:

$$I_C = \left[\frac{V}{Z}\right] \text{ Amps}$$

Where,

I_C = circulating current
V = induced voltage
Z = impedance of the sheath $\qquad Z = \sqrt{R^2 + (\omega M)^2} \quad \Omega / m$

13.2. Some practical situations

Note that the circulating current is independent of cable length and system voltage.

13.2.1. Short cable length

Earthing one end of the cable in this instance is recommended as the induced voltage will be very low, even under short circuit conditions.

13.2.2. Medium cable length

As the length of the cable run increases the induced voltage will also increase. It is quite possible that this voltage could be as high as 70 or 80 volts for a single drum length of cable. In the event of a lightning strike or system over voltage, this voltage would increase significantly.

One method of controlling the induced voltage in the event of an over voltage is to use surge arresters connected to the cable sheath, armours and the equipment earth. These arresters will be set to operate at a slightly higher level than the standing voltage under normal operating conditions.

13.2.3. Long cable runs

In this situation potentially lethal over voltages can be induced and it is necessary to reduce or eliminate totally the standing voltage. One of the simplest ways is to divide the cable run into a multiple of three equal sections and install cross bonding kits which are usually an additional part of a joint. The effect of cross bonding of the earth is to cancel, or significantly reduce the induced voltage. A typical arrangement is shown in fig. 110.

The practicalities of cross bonding are quite simple. An earth lead is taken from each side of the joint and these leads are then connected to each other in the sequence shown above.

Figure 110
Cross bonding of long single core cable run

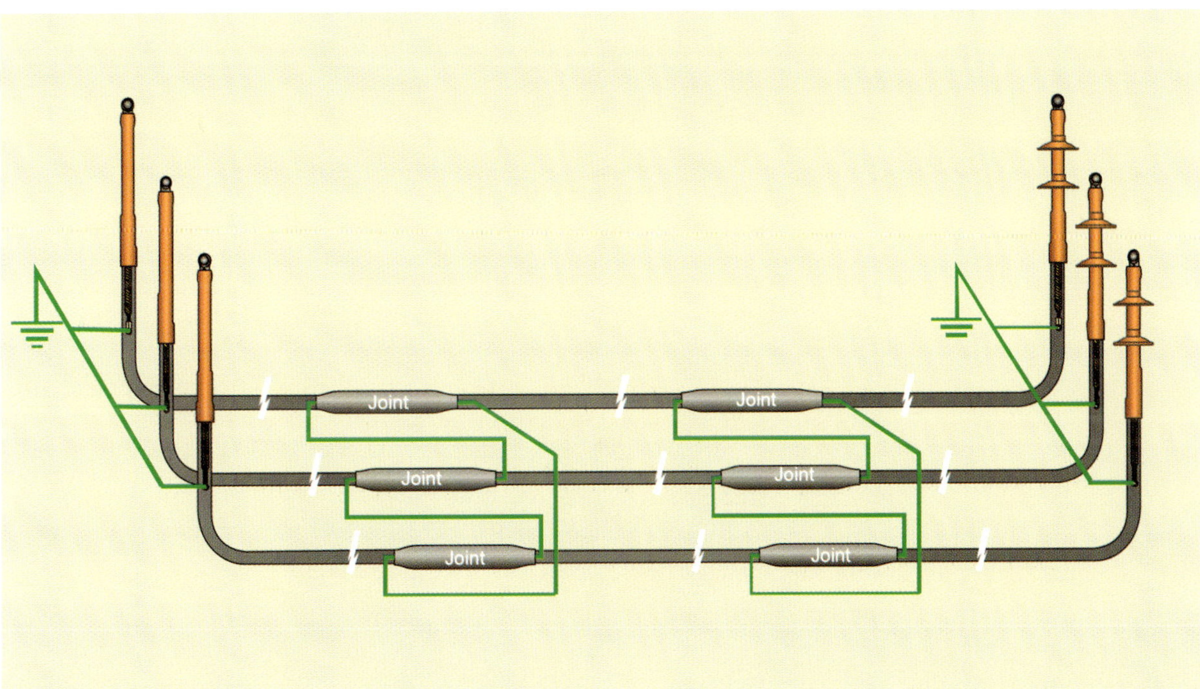

Figure 111
Failure due to
circulating currents

Figure 112
No problem at
opposite end

13.3. Fault caused by circulating currents

Figure 111 shows a failure due to circulating currents. Figure 112 shows the same cable at the opposite end which shows no sign of distress.

The failure was due to a poor earth connection at one end of the cable run which was earthed at both ends. It was later found that the circulating current was 70 A which the end user was happy with but this required suitable earthing. Unfortunately the connection was poor and overheating eventually lead to breakdown of the cable.

The opposite end of the cable run was terminated into a top entry cable box fitted to switchgear. The glands used to earth the aluminium armour of the cables were of the type shown in fig. 69. They were perfectly made off and were not in the least distressed by the circulating current.

13.4. Conclusions

It is obvious that the earthing arrangement of single core cables should be carefully considered. In the above example, if a good earth connection was made at the end which failed then there would not have been a problem. Whether a 70 A circulating current is acceptable to everyone is another matter, especially as this leads to the de-rating of the cable current carrying capacity.

It should be remembered that the circulating currents induced are independent of cable length, thus a short link between a transformer and unit of switchgear will be the same as a cable run of several hundred metres. In the former case earthing at one end is preferable since there will not be high induced voltages at the open end, even under short circuit conditions. In the second case a short circuit fault could lead to several hundred volts being induced, which could pose a possible danger to human life or flashover between the unearthed armours and the equipment earth.

One further point. If earthing at one end it should be the supply end.

Fortunately if the potential problems associated with single core earthing are known, solutions are available.

Chapter 14

Type testing of accessories

14.1. Introduction

Cable accessories are the only component in a distribution network that cannot be factory tested prior to service because they are assembled on site (with the exception of screened connectors and the testing of these products will be treated separately in section 14.6.) The only test they are subjected to after installation is a short-term DC pressure test, which will only show up the most fundamental errors of a jointer. Such an error would be the failure to remove the earth screen from the cable core – something the author has witnessed on more than one occasion. A correctly prepared cable would probably pass the pressure test without a termination fitted, provided the person carrying out the test ignored the noise of the resultant corona.

Since dry accessories were introduced, there has been extensive work carried out to find suitable test programs to prove a product is suitable for use in a given system. Many national specifications were produced and generally contained similar tests, but with differing requirements regarding test levels and duration. The subject is now addressed at an international level and two principal bodies for issuing specifications, IEC (International Electrotechnical Committee), CENELEC (European Committee for Electrotechnical Standardisation) have brought together some harmonisation. CENELEC have published a specification which embraces all the tests deemed necessary to evaluate any dry type accessory, be it heat shrink, cold shrink or push-on, including screened elbows. This specification is CENELEC HD 629.1 S1: 1996 for polymeric insulated cables and CENELEC HD 629.2 for paper cables. These specifications define all tests and test levels required for voltages up to 20.8/36 (42) kV. Certain countries have issued this specification under their national specification body as, for example, in the United Kingdom where it is included in BS 7888. The test methods are specified in CENELEC HD 628 S1: 1996.

The equivalent American specifications are IEEE 48 for terminations (although this does cover the range from 2.5 kV through to 765 kV) and for joints, IEEE 404.

Most European countries now use the CENELEC specification as a basis for their approvals for cable accessories, but many utilities insist on additional tests, which they consider relevant to their situation. However, it should be remembered that there is no substitute for field experience and that some of the issues addressed in the previous chapters are not covered by these tests. In addition, tests carried out in laboratories are performed on samples made under ideal conditions by skilled personnel. In the field the installers may not be well skilled and conditions may be far from ideal. As a result of this manufacturers are constantly striving to make their accessories easier to install and eliminate problems due to installer error.

It should be noted that the CENELEC specification is, at the time of writing, under review and the following sections may no longer be relevant. However, there is unlikely to be any major changes.

14.2. Some qualification tests for accessories

The following sections discuss the tests found in CENELEC HD 629.1 S1: 1996 and will give an indication of the tests used to evaluate dry type cable accessories. HD 629.2 for paper cable accessories is similar, but does not have a partial discharge requirement because of the inherent properties of the cable. A table at the end of this chapter will give the test voltage levels for each test along with the order.

A typical test arrangement for single core accessories is shown in fig. 113. It is necessary to test four joints and four terminations. For three core cables the connections are similar, but for qualification purposes two joints, which can be in series, are required thus one cable loop is sufficient.

It will be seen in fig. 113 that a third cable is used and this is to facilitate the installation of the current transformer, especially during the load cycling under water.

Figure 113 Typical test arrangement for single core cable accessories

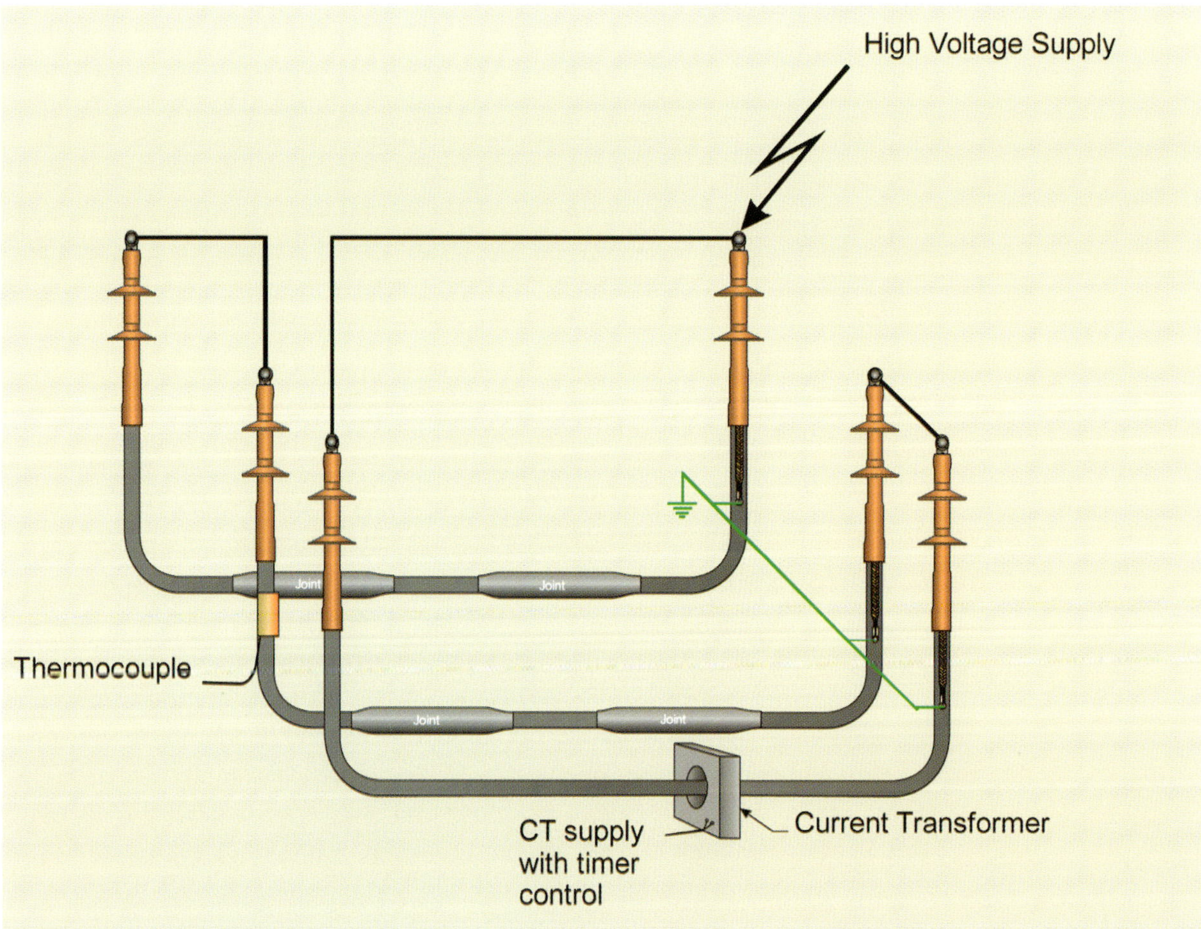

This cable will often be of the largest cross section of any tests to be performed and can therefore be used repeatedly. It also gives the manufacturer experience of much longer testing of accessories than is experienced in the normal test.

When testing screened cables it is sufficient to use a single phase transformer to apply the test voltage, but on three core paper belted cables a three phase transformer must be used.

The cable loops which contain joints will be required to be immersed in a tank of water deep enough to cover the joints to a depth of 1 metre. Joints can be immersed in pressure vessels containing water, which covers the joints, to simulate a depth of 1 meter. Should testing to simulate greater depths be required, the pressure vessel will be able to achieve this quite easily.

14.2.1. DC withstand

This is a 15 minutes test at 6 U_0 and is simply an initial pressure test. There is some debate as to the value of this test immediately preceding the AC test, but it may show up potential weaknesses in the samples.

14.2.2. AC dry withstand

These tests are usually performed near the beginning and at the end of a test program. Initially they are carried out at 4.5 U_0 and primarily they will check that nothing untoward exists with the termination prior to embarking on the full test program.

This test will highlight any poorly constructed samples in the first instance prior to the commencement of the full test program and will also show if there has been excessive deterioration of the product during testing when the final test is carried out. A failure at the start will indicate a serious problem, which may or may not be easily rectified. The installer forgetting to install the stress control for example is not unheard of, but of course is easily rectified.

14.2.3. AC wet withstand – outdoor terminations only

The wet test is performed on outdoor terminations and will only show a problem if the termination is too short. It is not normally a difficult test to pass, however, a problem could arise if the termination contains several sheds spaced closely together in order to reduce the overall length whilst maintaining the required creepage distance. In this instance water will fall between the sheds and effectively short them out.

14.2.4. Partial discharge

It should be pointed out that this test is not relevant on paper cable terminations because the cable will probably be discharging quite heavily at working voltage. This test checks that there are no significant discharges occurring and is one of the most meaningful tests on polymeric cable. It is usually carried out at 2 U_0 and the level must be below 10 pico coulombs. This test level has recently been increased from 1.75 U_0. In practice, a good accessory will show no significant discharge (i.e. < 1 pC) at the test voltage. Should discharging start below the test voltage and increase as the voltage increases even though it is within specification, it is the author's opinion that further tests on other samples should be performed to see if this is a function of the product. If the product is suspect under laboratory conditions, it is going to be especially vulnerable to field installation.

Provided the partial discharge test is satisfactory, the test program will proceed and further partial discharge tests will be performed throughout the program. Some of these tests will be carried out at high temperature – typically 5 to 10 °C above the rating of the cable, which for XLPE insulation is a maximum of 100 °C. This is to check that there is no change in the performance of the stress control system with heating and cooling of the termination.

14.2.5. Impact test – joints only

This test is carried out on joints only and involves dropping a weight onto the joint to check that it is mechanically sound. The weight will be dropped over the connector area and also over the sheath cut back point. This test will indicate a joint's ability to withstand impacts such as those caused during the back fill process on site.

14.2.6. Thermal cycling

This is the one long-term test and will last continuously for six weeks. The cable and accessories are heated to about 100 °C for a period of time and then allowed to cool whilst a continuous voltage of 2.5 U_0 is applied. A typical circuit is shown in fig. 113.

It is necessary to carry out tests on a similar sample prior to beginning the full test cycle to determine the current required to heat the conductor to the required temperature. A hole will be drilled through the cable to allow a thermocouple to be in contact with the conductor and a second thermocouple attached to the cable sheath. The heating and cooling cycle is specified as five hours heating (the cable should have reached its test temperature within three hours of the start of the cycle)

Figure 114
Typical thermal cycle for CENELEC test

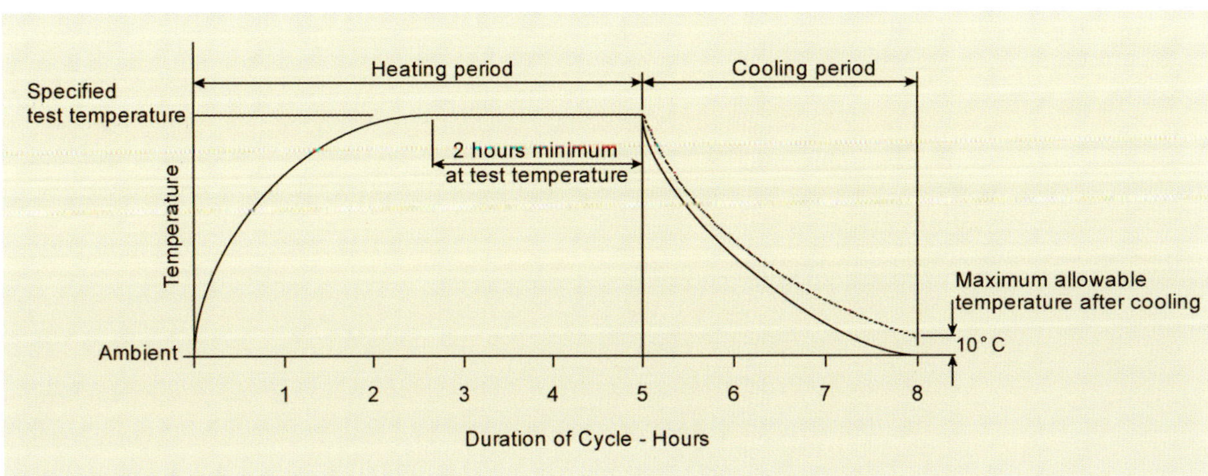

and three hours cooling. At the end of the cooling period the temperature should have fallen to within 10 °C of ambient. A typical thermal cycle is shown in fig. 114.

The current required to achieve the correct temperature is determined by trial and error and when the test conditions are met the temperature of the sheath is noted. This will then be used in the test to provide a continuous print out of the cycles. It is possible to apply a thermocouple to the conductor when the cable is heated prior to the application of voltage because the current transformer injects the current at very low voltage. Of course once the test program commences and voltage is applied it is no longer possible to do this, hence the reason for the preliminary test to determine the sheath temperature.

The test specified in IEEE 48 is shorter in duration but the test temperature is higher – 130 °C. Each cycle lasts for 24 hours of which the heating cycle is long enough to reach the test temperature and then hold it for six hours. The remainder of the cycle is one of cooling. A total of 30 cycles constitutes the test.

This test is designed to replicate service conditions where the cable temperature fluctuates due to the load variations throughout a given period. The effects of expansion and contraction of the cable plus the effect of temperature on the components are assessed. Partial discharge tests will indicate if there is significant change in the termination.

14.2.7. Impulse voltage withstand

The impulse test checks the accessory's ability to withstand a sudden over voltage. The waveform of the impulse is typically 1.2/50μ seconds which means the maximum voltage is reached in 1.2μ seconds and it then falls to 50% of its maximum 50μ seconds after the start of the test as shown in fig. 115. This waveform is representative of a lightning strike on the system and is normally a simple test to pass, especially on a termination which is not connected to other equipment. For system voltages above 17.5 kV it may be necessary to incorporate rain sheds on a termination to achieve this level, but otherwise there is no serious problem.

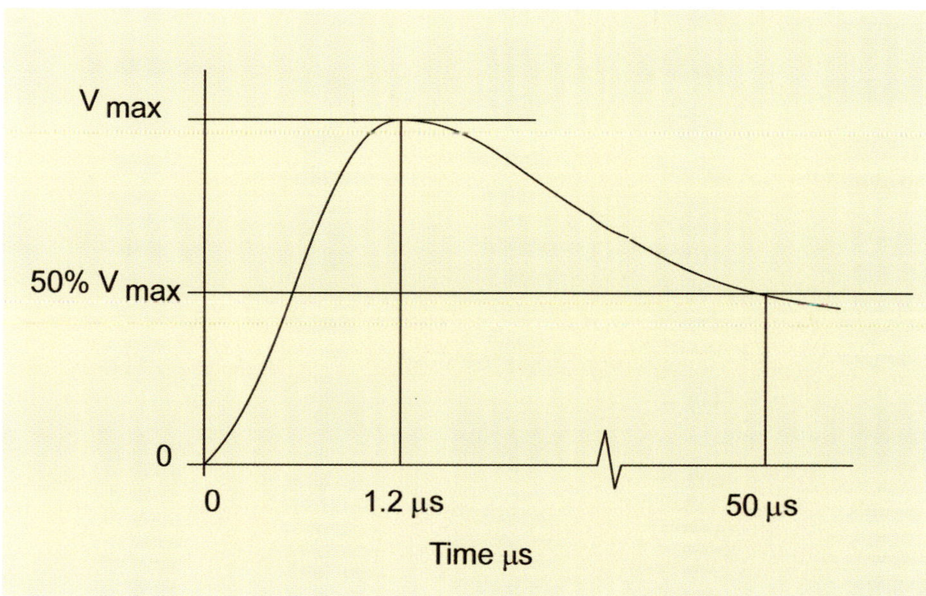

Figure 115
Typical impulse waveform

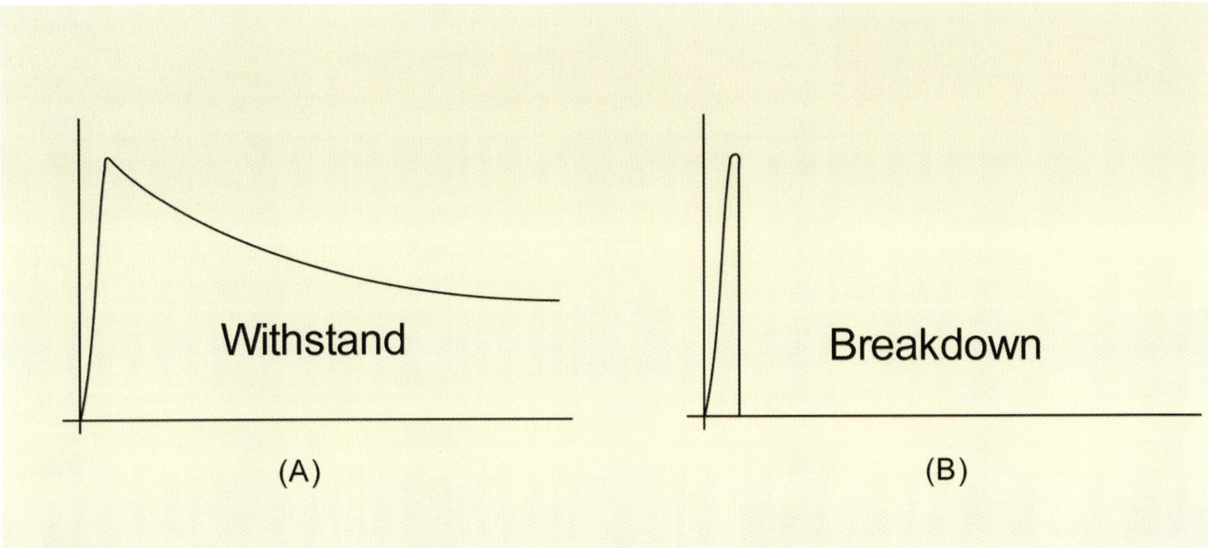

Figure 116
Typical waveforms obtained during testing

Figure 116 (A) shows the wave obtained if breakdown does not occur. Figure 116 (B) indicates when there is a breakdown. If the breakdown occurs over the surface of a termination then the failure is obvious from the visible flashover and accompanied loud noise. An internal failure of a joint or termination is not obvious, hence the need to record each waveform.

When the termination is connected to a switchgear bushing, the test is often more difficult to pass. The problem arises because the size of the bushings and their spacing may make it necessary to use stress relieving tape over the metal connection in addition to an insulating boot to achieve the necessary withstand. The impulse requirement for a 12 kV system is 75 kV and this rarely causes a problem, but many switchgear manufacturers now use the same bushing for 17.5 kV where the impulse level is 95 kV. The problem is made more difficult because there are so many different bushing and cable box designs that testing a termination on every possible piece of equipment is logistically virtually impossible. The one saving grace is that the possibility of such an over voltage occurring in practice and causing a failure is practically non-existent. There have certainly been no reported failures on indoor terminations resulting from such an over voltage in the author's experience.

Manufacturers qualifying a termination will usually carry out the tests on two or three representative bushings and cable boxes, which are known to be close to the design limit.

14.2.8. Humidity – indoor terminations only

This test is used to ascertain an indoor termination's ability to operate in a humid environment. The test involves placing the termination in a chamber fitted with nozzles, which create a mist that can settle on the termination, which is energised at 1.25 U_0. The duration of the test is 300 hours and it will assess the outer materials' ability to withstand the effects of partial discharges. It will also test the end seals against moisture ingress. The test voltage is relatively low because it is intended to simulate the practical case. Much higher test voltages may be less onerous as the surface discharges are more active and mobile and therefore less damaging than those which root on one spot.

14.2.9. Salt fog – outdoor terminations only

This test is for outdoor terminations and can be carried out using the same apparatus as for the humidity test. The difference is that the spray, which creates the mist, is far more conductive and surface discharge activity is much greater. This test lasts 1000 hours with an applied voltage of 1.25 U_0. At the end of the test the termination is examined for signs of tracking and severe erosion – either of which is not acceptable.

14.2.10. Immersion test – outdoor terminations only

Terminations are immersed in water to check the water seals. For single core cables this is simply the seal onto the lug at one end and onto the sheath at the opposite end. In the case of three core cables the sealing properties around the cable breakout are also tested.

This test involves immersing the whole of the termination in water to a depth of at least 0.03 m and performing thermal cycles as described above. In this test, no voltage is applied for obvious reasons.

14.2.11. Thermal short circuit

Short circuit tests subject the cable to high temperature and in the case of three core cable, extremely high mechanical forces. The forces exerted on a single core system will require the cables themselves to be firmly anchored and this can be achieved using a cleat of the types shown in figs. 125 to 129 to hold each cable, but the accessories should not be affected. The thermal short circuit involves heating the cable conductor to either 160 °C for paper cable or 250 °C for XLPE cable. Equations are stated within the specification, which give the value of I^2t as a function of cable conductor material and cross section, ambient temperature and final test temperature.

The duration of the short circuit will depend upon the users' system and requirements and the cable size and can vary from less than a second up to a maximum of 5 seconds. The point of this test is to check the mechanical integrity of a three core accessory and the overall ability of both single and three core accessories to withstand a sudden increase in temperature to a very high level without causing irreparable damage to the accessory. When testing the metallic screen, loss of conductivity is checked at the various connections within the accessory. Any significant increase in contact resistance is not acceptable.

It should be mentioned at this point that earth connections and components should have undergone extensive load cycling and short circuit testing in their own right prior to being utilised in a joint or termination and therefore should not fail the short circuit test.

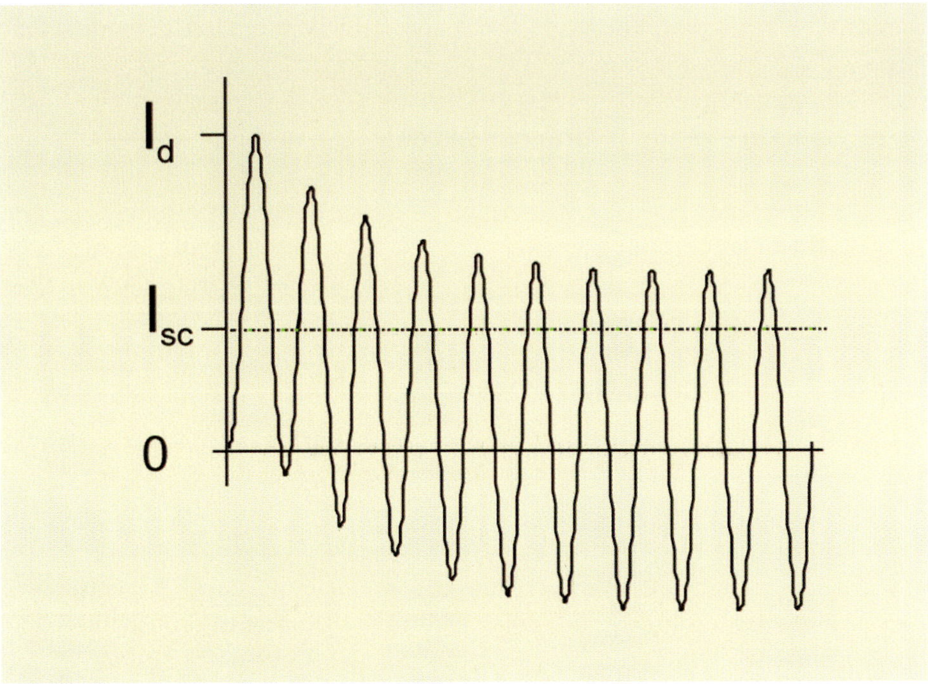

Figure 117
Typical waveform of one phase for short circuit test

14.2.12. Dynamic short circuit

This test is mechanically very severe because the first peak of current, I_d, will be 2.5 times the thermal short-circuit current, I_{sc}, required to heat the cable to its test temperature in one second (see fig. 117). The waveform shown is for one phase only. The other two phases will have lower peaks on the first cycle and point on wave switching is required to ensure the test level is achieved.

The accessory most likely to experience difficulty with this test is the three core termination where the breakout is likely to be stretched and can split. For high dynamic currents it may be necessary to reinforce this area of the termination. The duration of this test will be long enough to ensure that the first peak is reached and typically will last for at least 10 ms. The subsequent duration of the test may be such as to combine the thermal test also, although the two are often performed separately.

The short circuit tests can be performed on different cable loops to the remainder of the tests with AC and DC withstand performed before the short circuits and impulse and AC withstand tests carried out afterwards.

14.2.13. Inspection

At the conclusion of the tests the accessory will be examined closely and this entails completely dismantling it very carefully in order not to destroy any signs of degradation. The inspection should be looking for the following:

- Signs of tracking on terminations.
- Signs of severe erosion on terminations – this is rather subjective and relies on the experience of the witnessing engineer.

- Ingress of moisture – this could have occurred but not had time to effect the electrical properties.
- Seal under interfaces such as cable sheath/outer cover.
- Seal under rain sheds.
- Mechanical damage after short circuit.

14.3. Test Voltage Levels

The following table gives the voltage test levels specified by CENELEC for rated voltages from 12 to 36 kV. Complete test details can be obtained from the specification.

Tests	System Voltage kV				
	7.2	12	17.5	24	36
DC dry withstand	25	42	60	85	125
Initial AC dry withstand	20	30	45	60	95
AC wet withstand	20	30	45	60	95
Partial discharge	8.5	14	20	28	41
Impulse withstand indoor terminations	60	75	95	125	170
Impulse withstand outdoor terminations and joints	70	95	110	150	200
Load cycling	10	17.5	25	35	50

Tests	System Voltage kV				
	7.2	12	17.5	24	36
Humidity	5.2	8.6	12.6	17	26
Salt fog	5.2	8.6	12.6	17	26
Final AC withstand	10	17.5	25	35	50

Voltage test levels for accessories tested to CENELEC HD 629.
The following tables list the test sequences for joints and terminations.

14.4. Testing terminations to HD 629/BS 7888

Test	Requirement
DC voltage dry withstand	*15 minutes at 6 U_0*
AC voltage dry withstand	*5 minutes at 4 U_0*
AC voltage wet withstand (outdoor)	*1 minute at 4 U_0*
Partial discharge at ambient temperature	*10 pC max at 2 U_0*
Impulse voltage at elevated temperature	*10 +ve 10 -ve*
DC voltage dry withstand	*15 minutes at 6 U_0*
AC voltage dry withstand	*5 minutes at 4 U_0*
AC voltage wet withstand (outdoor)	*1 minute at 4 U_0*
Partial discharge at ambient temperature	*10 pC max at 2 U_0*
Impulse voltage at elevated temperature	*10 +ve 10 -ve*

Test	Requirement
Electrical heat cycling – 5 hours on 3 hours off (95°C max)	*3 cycles at 2.5 U_0*
Partial discharge at ambient and elevated temperature	*10 pC max at 2 U_0*
Immersion (outdoor only)	*10 cycles – no voltage*
Electrical heat cycling – 5 hours on 3 hours off (95°C max)	*123 cycles at 2.5 U_0*
Partial discharge at ambient and elevated temperature	*10 pC max at 2 U_0*
Thermal short circuit (different sample)	*2 short circuits*
Dynamic short circuit (same sample as for thermal)	*1 short circuit*
Impulse voltage at ambient temperature*	*10 +ve 10 -ve*
** 75, 95, 125, 170 kV for 12, 17.5, 24 & 36 kV*	
AC voltage dry withstand at ambient temperature	*15 minutes at 2.5 U_0*
Humidity (indoor) (separate sample)	*300 hours at 1.25 U_0*
Salt fog (outdoor) (separate sample)	*1000 hours at 1.25 U_0*
Examination	*For information*

14.5. Testing joints to HD 629/BS 7888

Test	Requirement
DC voltage dry withstand	*15 minutes at 6 U_0*
AC voltage dry withstand	*5 minutes at 4 U_0*
Partial discharge at ambient temperature	*10 pC max at 2 U_0*

Test	Requirement
Impact (optional and if needed at ambient and/or -20 ^{0}C)	*Insulation resistance >10^3 MW*
Impulse voltage at elevated temperature	*10 +ve 10 -ve*
Electrical heat cycling – 5 hours on 3 hours off (95 ^{0}C max)	*3 cycles at 2.5 U_0*
Partial discharge at ambient and elevated temperature	*10 pC max at 2 U_0*
Electrical heat cycling – 5 hours on 3 hours off (95 ^{0}C max)	*60 cycles at 2.5 U_0*
Electrical heat cycling –as above but under 1m water	*63 cycles at 2.5 U_0*
Partial discharge at ambient and elevated temperature	*10 pC max at 2 U_0*
Thermal short circuit (different sample) – screen	*2 short circuits*
Thermal short circuit (different sample) – conductor	*2 short circuits*
Dynamic short circuit (same sample as for thermal)	*10 +ve 10 -ve*
Impulse voltage at ambient temperature*	*10 +ve 10 -ve*
** 75, 95, 125, 170 kV for 12, 17.5, 24 & 36 kV*	
AC voltage withstand at ambient temperature	*15 minutes at 2.5 U_0*
Examination	For information

14.6. Testing Screened Separable Connectors

Test	Requirement
AC voltage dry withstand	*5 minutes at 4 U_0*
Partial discharge at ambient temperature	*10 pC max at 2 U_0*
Impulse voltage at elevated temperature	*10 +ve 10 -ve*
Electrical heat cycling – 5 hours on 3 hours off (95^{0}C max)	*3 cycles at 2.5 U_0*
Partial discharge at ambient and elevated temperature	*10 pC max at 2 U_0*
Electrical heat cycling – 5 hours on 3 hours off (95^{0}C max)	*123 cycles at 2.5 U_0*
Partial discharge at ambient and elevated temperature	*10 pC max at 2 U_0*
Thermal short circuit (different sample)	*2 short circuits*
Dynamic short circuit (same sample as for thermal)	*1 short circuit*
Impulse voltage at ambient temperature*	*10 +ve 10 -ve*
** 75, 95, 125, 170 kV for 12, 17.5, 24 & 36 kV*	
AC voltage dry withstand at ambient temperature	*15 minutes at 2.5 U_0*
Examination	*For information*

14.7. Limitations of type testing

Whilst the above series of tests are searching, they do have certain limitations as follows:

- *When type testing the cable is professionally prepared with correct tools.*
- *Accessories are installed in ideal conditions by qualified personnel.*

- *Many problems in the field arise from poor cable preparation and poor installation.*

- *The majority of faults not associated with poor cable prep and installation are due to factors not covered by the specification (this does not usually apply to separable connectors). For example:*

- *Indoor terminations are not humidity tested in a realistic environment, i.e. a cable box.*

- *Terminations are not tested in core crossing configurations.*

- *Tests carried out at elevated voltages may result in different breakdown processes to those in practice at rated voltage.*

Many authorities, aware of some of the limitations, call for additional testing which is relevant to their working environment.

Chapter 15

Surge arresters

15.1. Introduction

Whilst surge arresters are not classed as cable accessories they are increasingly used in parallel with separable connectors to provide additional protection against over voltages that occur on a system. It has already been mentioned in chapter 13, section 13.2.2., that the use of surge arresters will overcome the problem of high induced voltages when single core cables earthed at one end experience a short circuit. This chapter gives a brief overview of these arresters and their application in conjunction with separable connectors.

15.2. Metal Oxide Arrester

This type of arrester consists of a series of non linear resistive discs. These discs are voltage sensitive semiconductors which at normal voltage are insulators but above a certain voltage become conductive. This allows over currents to be diverted to earth thus protecting the associated equipment. Figure 118 shows a range of typical discs.

Figure 118
Metal oxide discs

Figure 119 shows how a surge arrester is constructed in a manner which enables it to be connected to a separable connector.

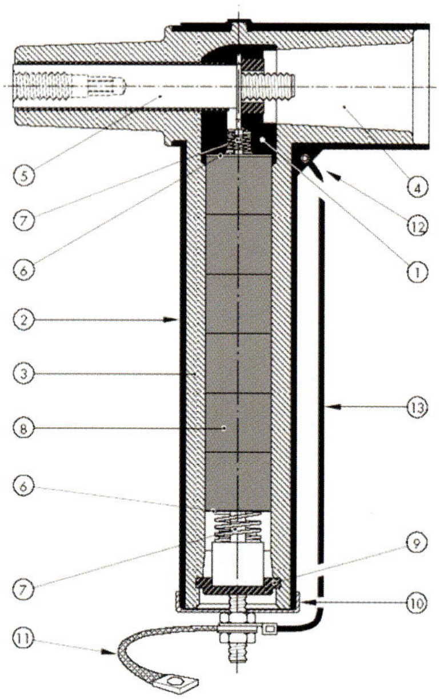

1. Faraday cage
2. Conductive outer jacket
3. EPDM insulation between elements and outer jacket
4. Interference fit to interface
5. Contact rod
6. Contact disc
7. Copper current shunt to eliminate inductive effects of strings
8. Metal oxide discs
9. Moisture seal
10. Stainless steel end cap
11. Earth connection
12. Earthing eye
13. Earth lead

Figure 119
Screened surge arrester

15.3. Typical surge arresters and connections

The following figures show typical surge arresters and connections to separable connectors.

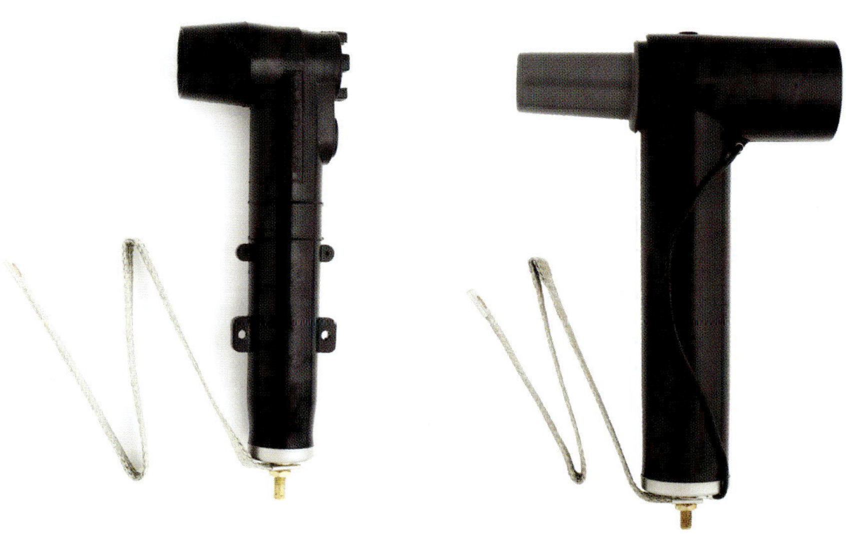

Figure 120
Typical surge arresters

Typical applications and dimensions

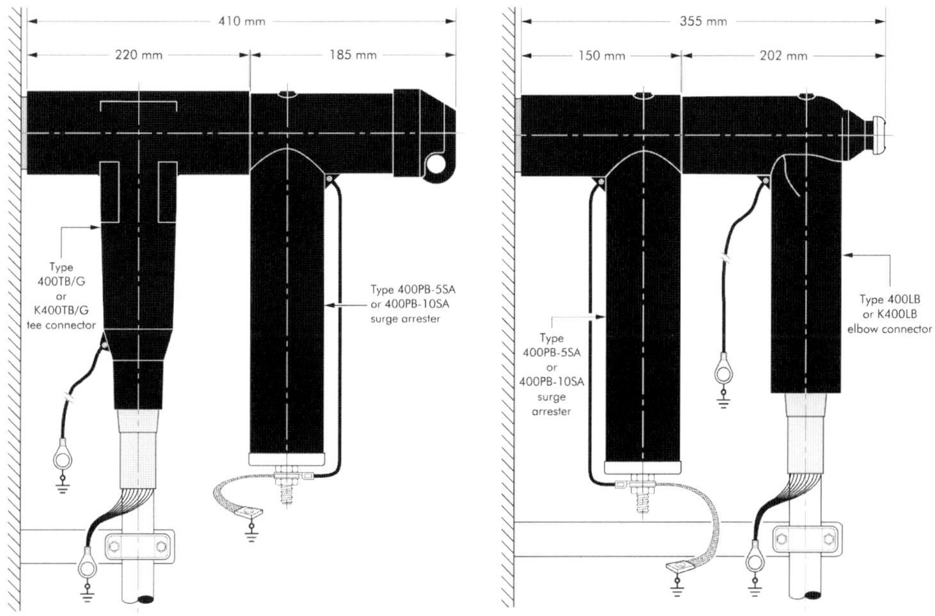

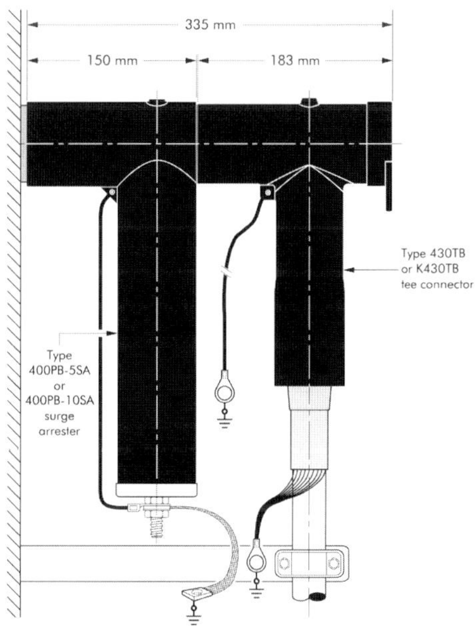

Figure 121
Typical surge arrester connections

Ordering instructions

To order the surge arrester, specify the surge arrester type, as described on previous page.

Example:

For a maximum continuous operating voltage (rms) of 24 kV and a nominal discharge current of 10 kA.
Order a 400PB-10SA-24N surge arrester.

Chapter 16

Testing of cable and equipment

It is often desirable to test cables and or equipment after installation or for fault location. This is carried out with the cables connected or disconnected. The use of separable connectors to connect the cable to equipment makes this task quite straightforward and the following gives typical examples of performing this work. The products shown here are produced by Nexans Div. Euromold.

A. 400TB & 440TB connectors

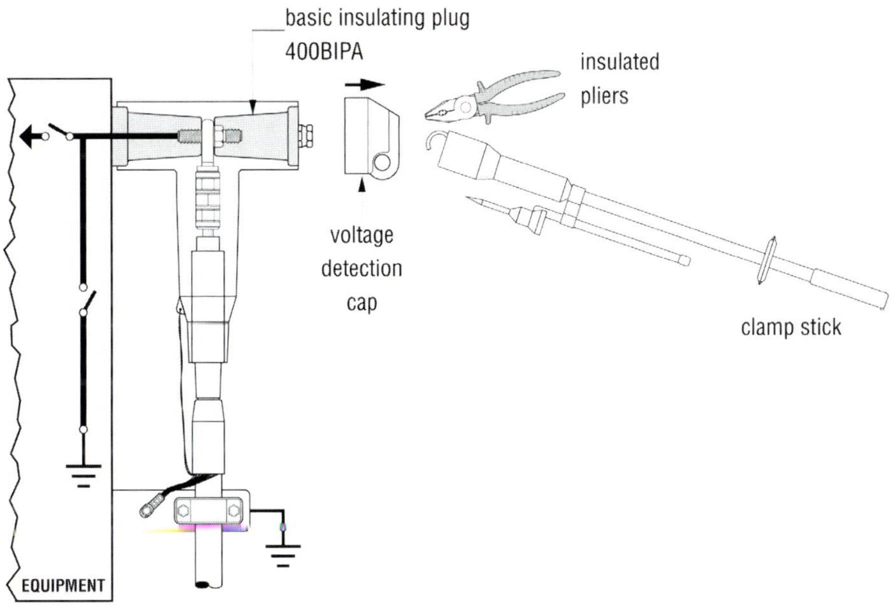

De-energise the equipment by opening the switches.
Padlock the switches to prevent operator error.
Insulated pliers or clamp stick are used to remove the voltage detection cap exposing the basic insulating plug.

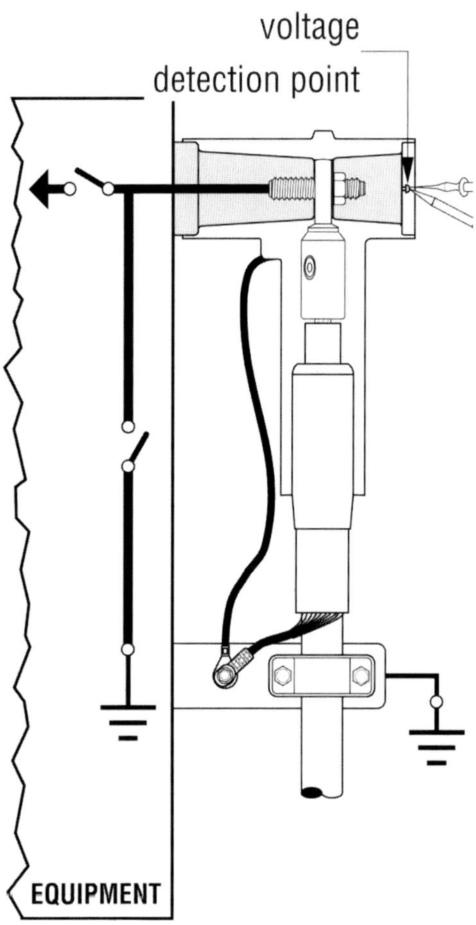

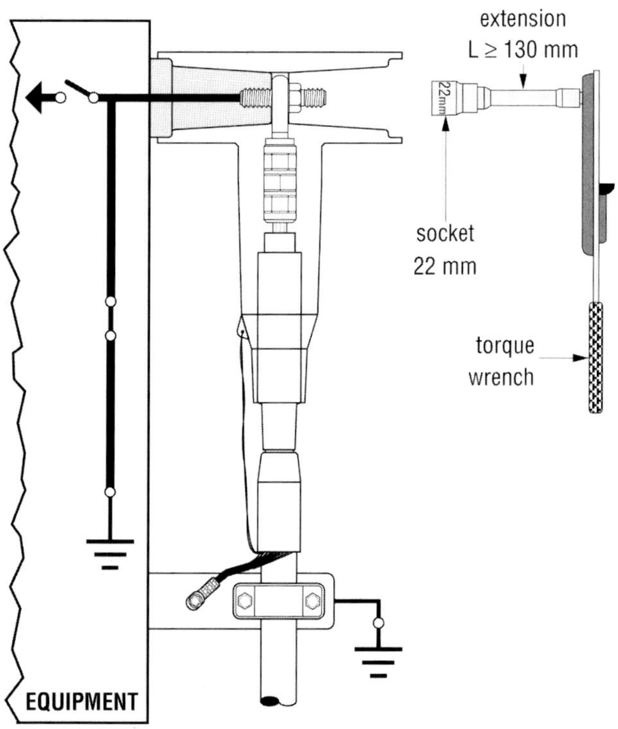

Make sure the clamping screw is still tight. If not, torque it to 50 Nm (a 22 mm socket, an extension and torque wrench is required).

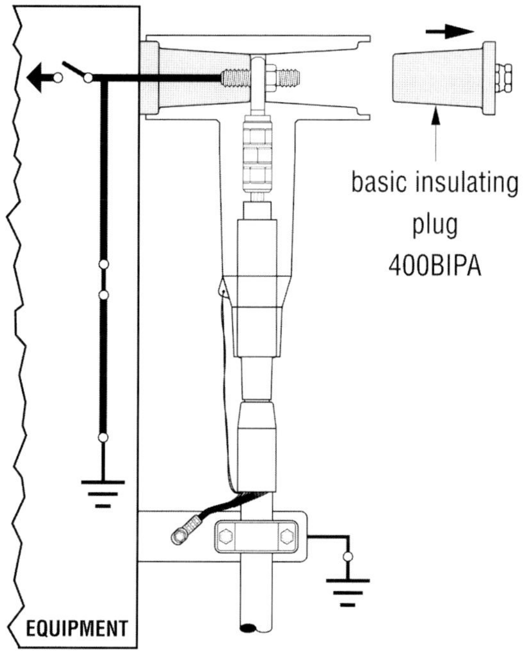

Once the cable has been tested and found de-energised, the basic insulating plug 400BIPA can be unscrewed and removed from connector with a 24 mm wrench.
Keep clean the basic insulating plug at all times.

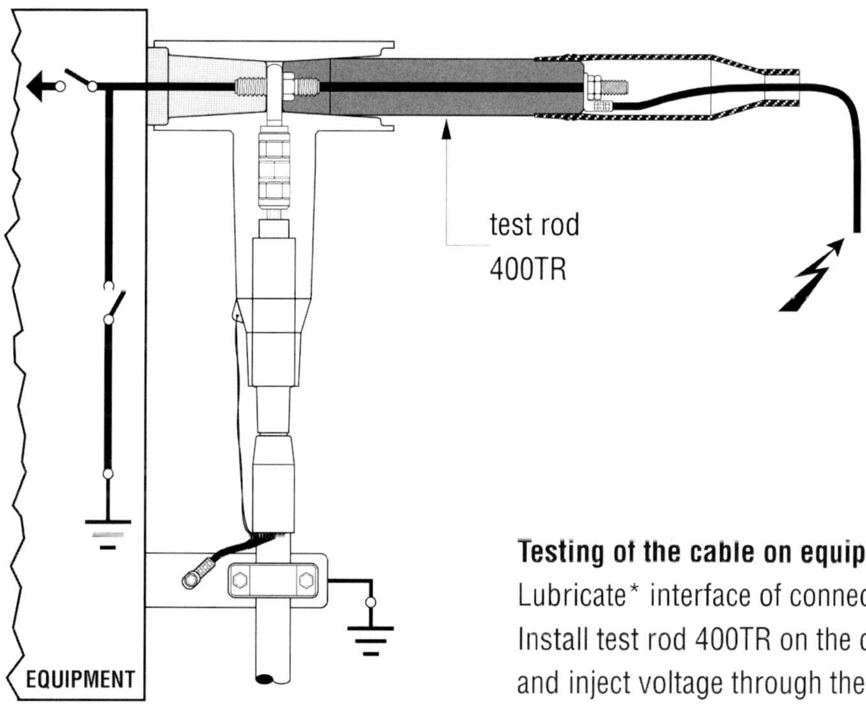

Testing of the cable on equipment :
Lubricate* interface of connector and test rod.
Install test rod 400TR on the connector (hand tighten) and inject voltage through the test rod

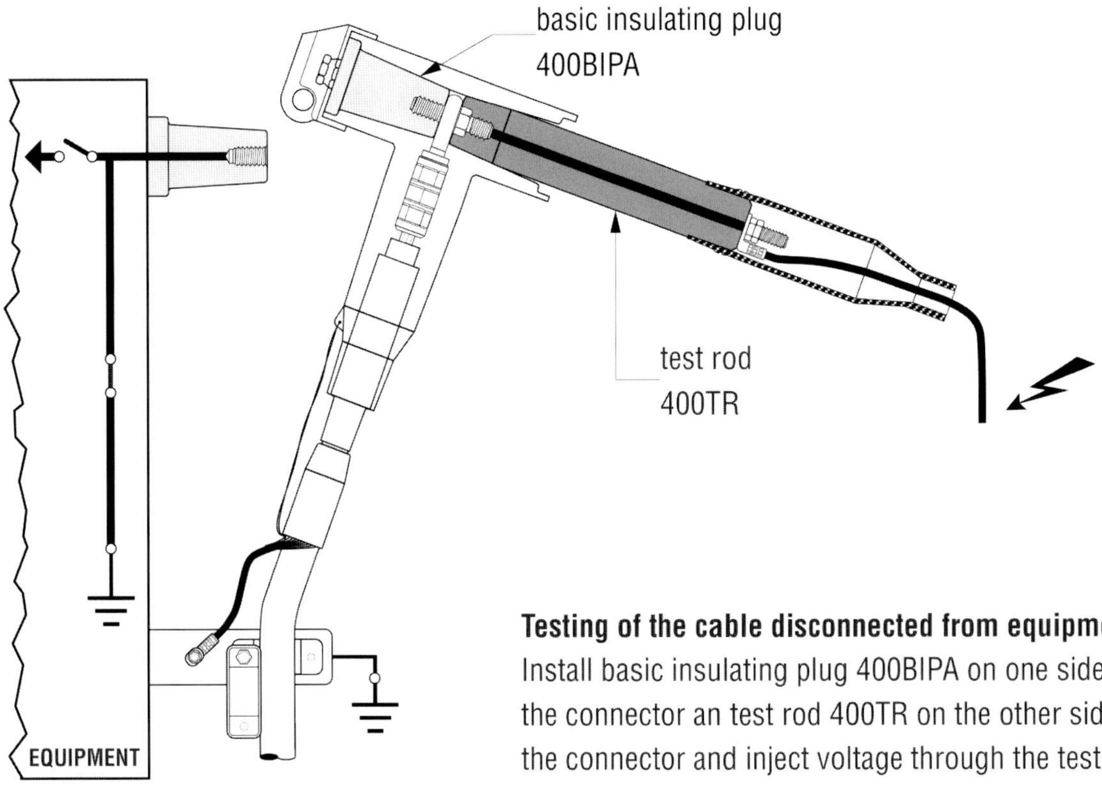

Testing of the cable disconnected from equipment :
Install basic insulating plug 400BIPA on one side of the connector an test rod 400TR on the other side of the connector and inject voltage through the test rod.

B. 430TB connectors

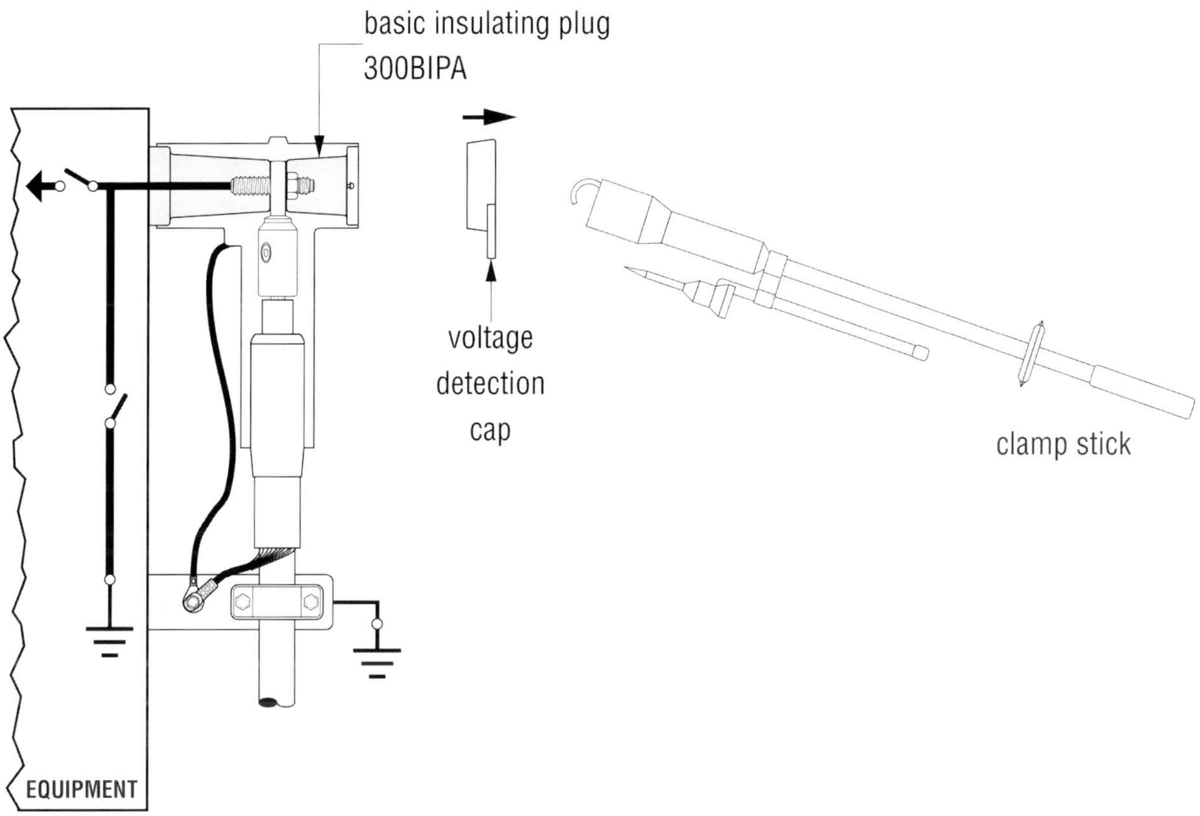

De-energise the equipment by opening the switches.
Padlock the switches to prevent operator error.
Insulated pliers or clamp stick are used to remove the voltage detection cap exposing the basic insulating plug.

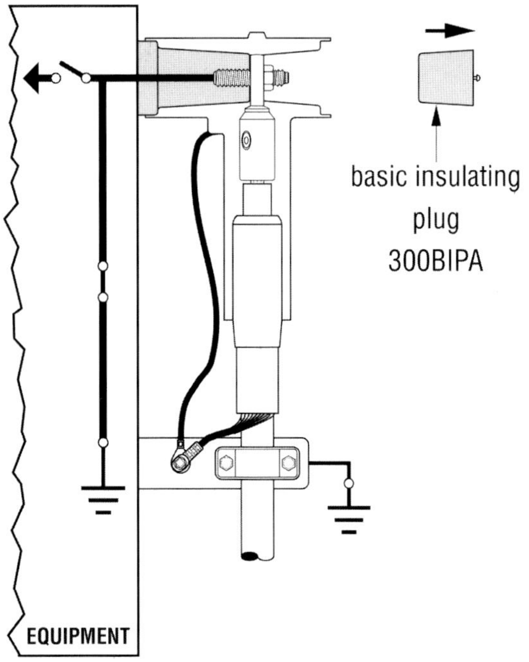

Once the cable has been tested and found de-energised, the basic insulating plug 300BIPA can be unscrewed and removed from connector with a 22 mm wrench.
Keep clean the basic insulating plug at all times.

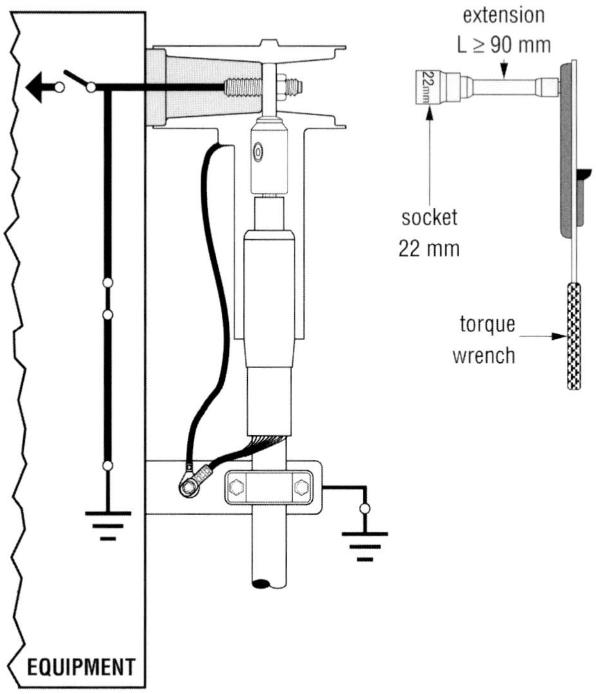

Make sure the clamping screw is still tight. If not, torque it to 50 Nm (a 22 mm socket, an extension and torque wrench is required).

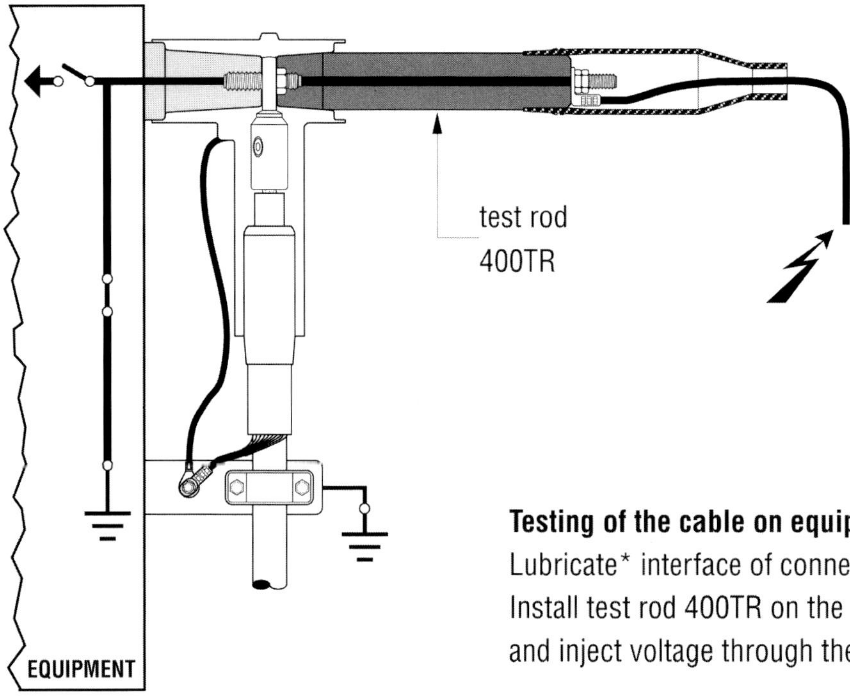

Testing of the cable on equipment :

Lubricate* interface of connector and test rod.
Install test rod 400TR on the connector (hand tighten) and inject voltage through the test rod

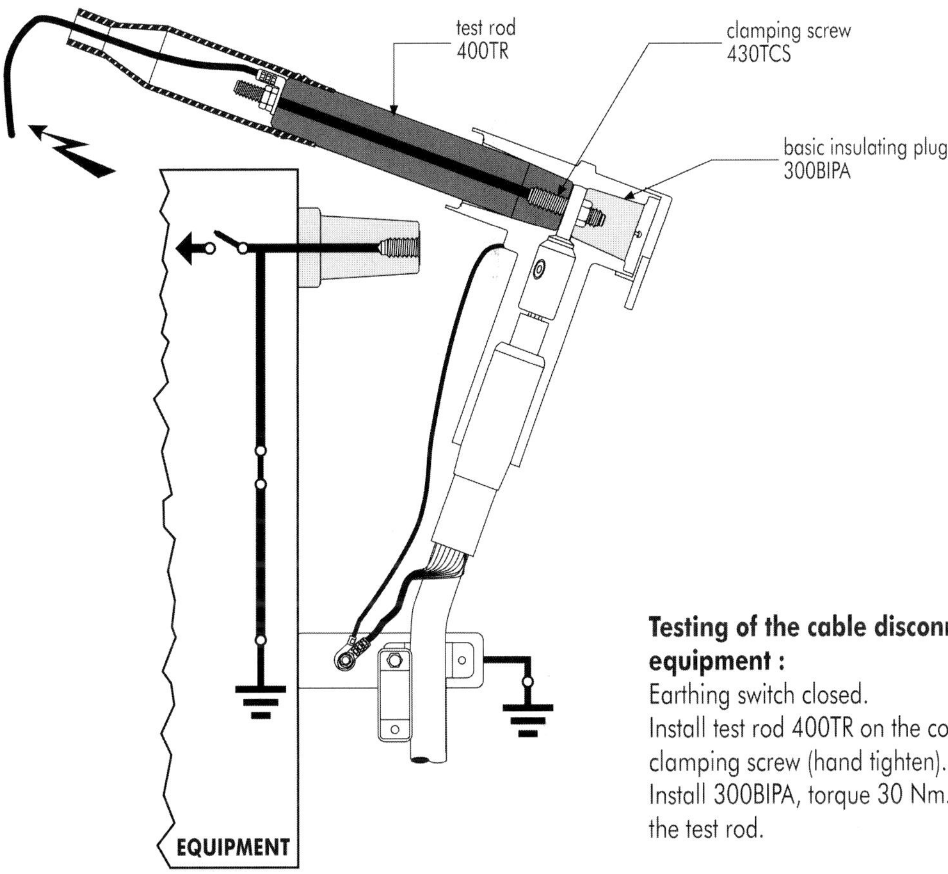

Testing of the cable disconnected from equipment :
Earthing switch closed.
Install test rod 400TR on the connector with the clamping screw (hand tighten).
Install 300BIPA, torque 30 Nm. Inject voltage through the test rod.

Chapter 17

Cable supports

Cables should always be adequately supported for the following reasons:

- Cable lugs used with terminations and separable connectors should not support the weight of the cable. Without support there is excessive bending force on the bushing and the cable could be pulled from the lugs or connectors.

- Similarly, joints fitted to cables in the vertical position could result in the cable being pulled from the connector. Even partial movement of the cable can result in void formation and internal discharges causing total breakdown.

Figure 122
Cable ties used on single core cables in trefoil

Figure 123
Short circuit test on single core cables with stainless steel cable ties

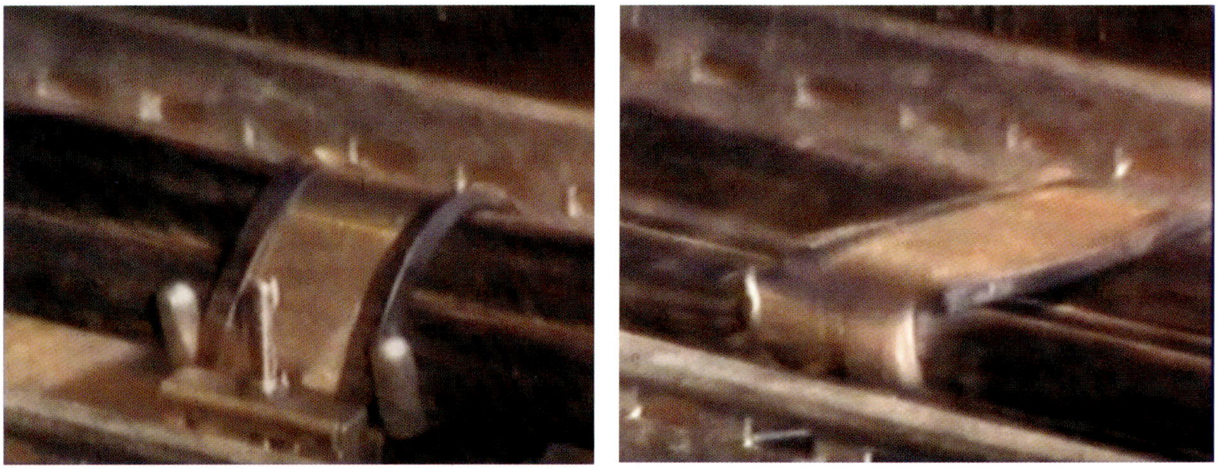

Figure 124
Short circuit test on single core cables with stainless steel strap

- Under short circuit conditions the resultant mechanical forces cause extreme movement between single core cables. To prevent severe mechanical damage the cable should be supported in such a manner that the movement is restricted.

It is not unheard of for plastic cable ties being used to hold single core cables in a trefoil formation – see fig. 122. Clearly this will give little support at any time and certainly not in the event of a short circuit. Even when using stainless steel cable ties, this method offers limited support under short circuit.

Figure 123 shows what happens to cables held with stainless steel cable ties during a short circuit test. Figure 124 indicates the testing of a substantial strap which also fails spectacularly shortly after the second photograph. The test duration was in the order of a few cycles.

It is imperative, therefore, to use a more robust system to prevent cable damage during a short circuit.

17.2. Cleats

Cable cleats are designed to support cables and prevent excessive cable movement under short circuit conditions. There are numerous cable cleat designs on the market and the following section will give a brief overview.

17.2.1. Requirements

In addition to the obvious need to restrain cable movement, the operating environment must be taken into account. When considering the choice of cleat for a given application it is best to discuss requirements with a reputable manufacturer who will talk through the issues involved to arrive at the best solution. The following information is required to arrive at the optimum solution:

- Cable diameter.
- Cable construction (e.g. armoured, unarmoured).
- Short circuit rating of the system (the cable used may be over specified).
- The environment.
- Cable run support (e.g. cable trays, hangers, etc).

17.2.2. Cleat spacing

Cleat spacing is a carefully calculated exercise and the specifier requires the follow information:

- The cable diameter (when single cables are in trefoil).
- The centre to centre distance (when the cables are laid in flat formation or when considering multi-core cable).
- The peak fault current in kA.
- The physical strength of the proposed cleat.

17.2.3. Cleat specifications

There are a large number of cleat designs available and they should be manufactured to meet one of the two specifications in existence as appropriate. These specifications are IEC61914 – 2009 and EN50368.

17.2.4. Trefoil cables with low short circuit levels

Figure 125 below shows such a product manufactured in aluminium. To prevent galvanic corrosion a polymeric base with low smoke and fume and zero halogen properties should be installed.

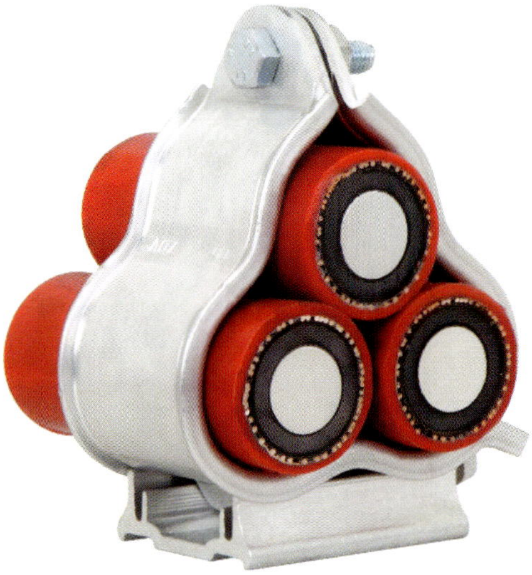

Figure 125
Aluminium cleat for single core cables in trefoil

17.2.5. Trefoil cables with moderate short circuit levels

As the short circuit level increases the forces generated also increase as does the temperature of the cable (the conductor can reach up to 250 °C). The resulting temperature rise leads to a softening of the cable insulation and outer sheath leaving it susceptible to damage by a cleat. To prevent this, the cleat will have an inner

Figure 126
Cleats with polymeric liner for moderate short circuit

polymeric liner as shown in fig. 126. In addition aluminium will not have sufficient mechanical strength to hold the cables and thus stainless steel is used.

This shows both trefoil single core cables and a four core low voltage armoured cable.

17.2.6. Trefoil cables with high short circuit levels

Figure 127 shows the ultimate cleat for high short circuit levels.

Figure 127
Cleats with polymeric liner for high short circuit

17.2.7. Triplex cables

The examples above have shown cables arranged in trefoil. However, with triplex cable the three cable cores are twisted around each other. This would make fitting cleats to these cables difficult as the relative position of the cable varies at intervals along the cable run.

To overcome this problem, the following solution is offered. An inner liner shown in fig. 128 is applied around the cables and then a standard cleat is fitted (fig. 129).

In the fig.129 (a) a polymeric cleat is used and this will be suitable for low short circuit rating. For high short circuits a stainless steel cleat would be used (fig.129 (b).

This chapter is merely an introduction to cleats. There are several other systems not covered here, but as always it is recommended that expert advice on product selection be obtained from the manufacturer.

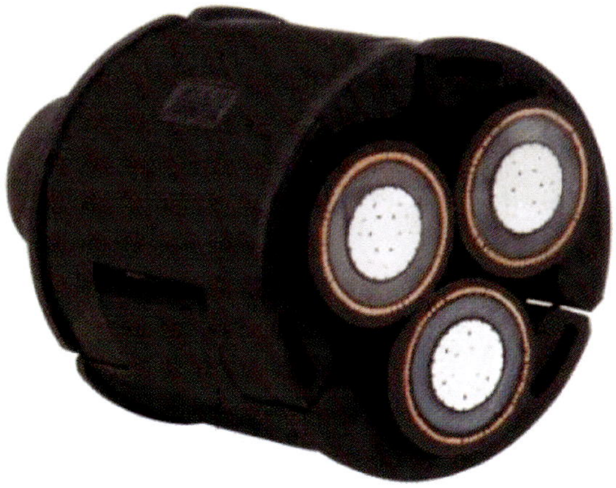

Figure 128
Triplex cable surround

Figure 129
Triplex cable surround inside standard cleats

17.3. Short circuit test with correctly installed cleats

When correctly specified cleats are installed the result of a short circuit is far less severe than shown in figs. 123 and 124. Figure 130 shows a trefoil cable arrangement after a short circuit test. It will be noticed that the cables are separated between the cleats but after the test the cables were electrically sound.

Figure 130
Before and after a short circuit test

Chapter 18

Recent developments

Single core cables are now much more popular than three core and as a result the accessory manufacturer can simplify the accessories offered whilst incorporating all the essential components required for satisfactory operation. The trend is to reduce the number of components used but it should be remembered that the greatest time taken to install an accessory is taken up with the preparation of the cable.

18.1. Terminations

The trend is to reduce the number of individual components used to make a termination. Some heat shrink products now incorporate a stress relieving material

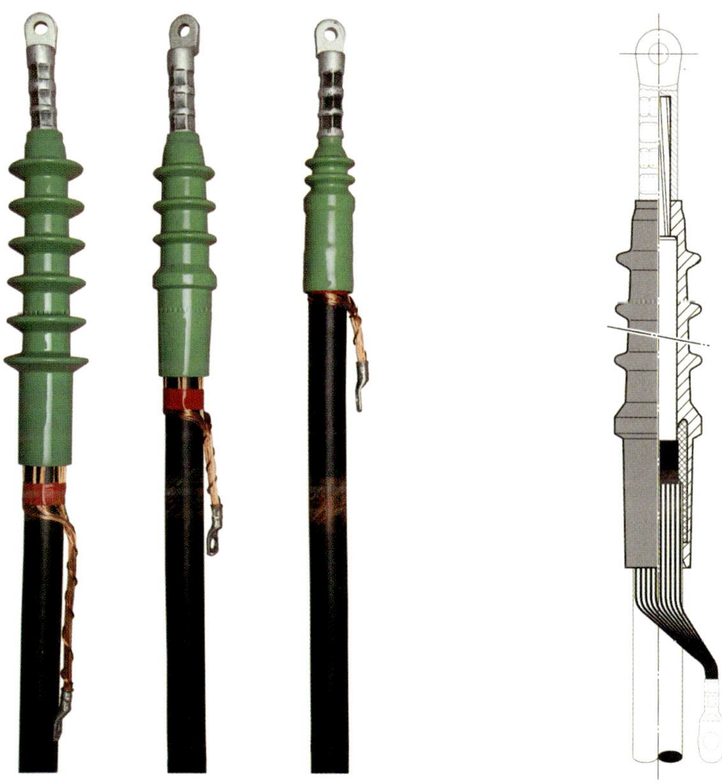

Figure 131
A push on termination

along the whole length of the outer tube which dispenses with the need for a separate stress control tube.

The main advances in cold applied products have been in the delivery system using a variety of hold outs.

18.1.1. Push on

Push on moulded products produced in silicone rubber enabling extensive range taking are a relatively recent addition. These have a moulded conductive cone included within the body of the termination as shown in fig. 131.

18.1.2. Cold shrink

Cold shrink terminations generally vary only in the hold out system. One point to bear in mind is that unlike heat shrink products; cold shrink does have a shelf life. If stored on a hold out for a long time it will not return to its original internal diameter. This is generally not a problem as there is little loss of memory for at least two years. Figure 132 shows a typical product on a holdout and then installed.

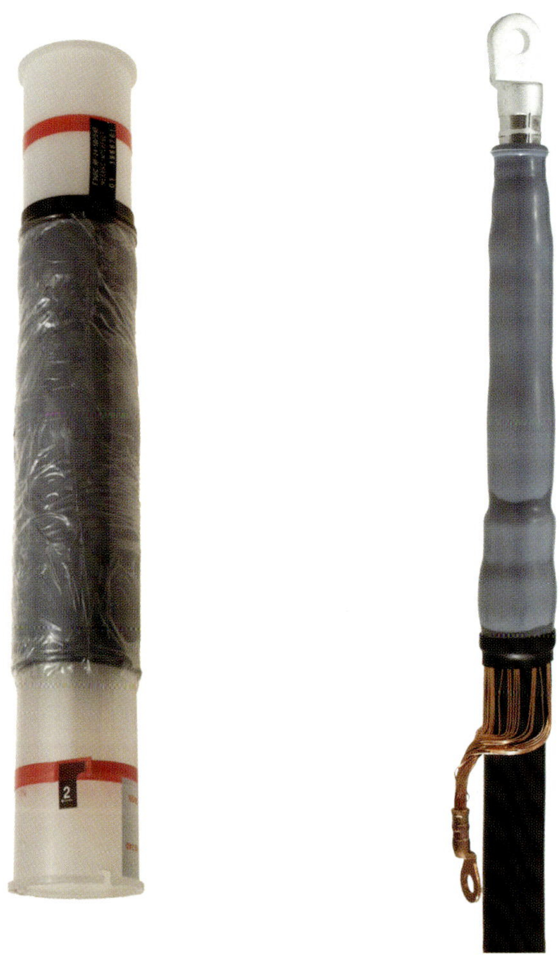

Figure 132
An indoor cold shrink termination

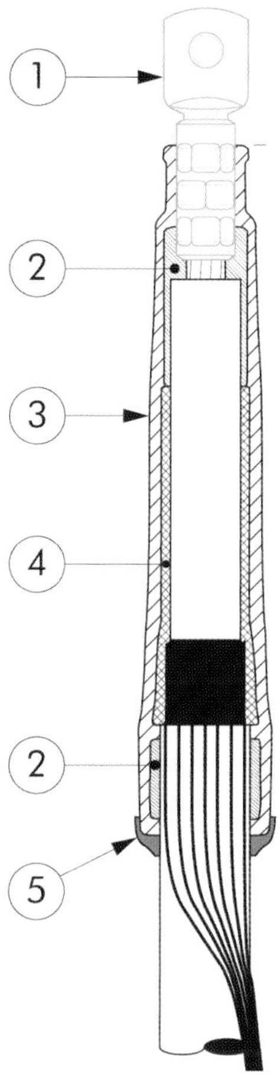

Indoor termination comprising

1. Cable lug (not included in the standard kit).
2. Water sealing mastic.
3. Silicone tube.
4. Stress control mastic.
5. Conductive EPDM ring.

Figure 133
An indoor cold shrink termination in section

Figure 133 shows the above termination in cross section. The stress control is pre- applied to the outer tube. For outdoor terminations rain sheds are moulded onto the outer tube. The conductive ring at the bottom of the termination is to ensure equal current flow under polluted conditions. Without this the current would flow from the termination surface to the earth wires leading to possible degradation of the silicone tube.

18.2. Joints

As with terminations, the move is to incorporate all the components into a single unit. Once more the delivery system is the main difference between different manufacturers. Figure 134 shows one of the latest such joints although some incorporate a Faraday cage. The various steps are self explanatory.

18.3. Separable connectors

18.3.1. Larger rated systems

Once again the design of these products has changed little since the product was introduced. The latest advance here is the introduction of a larger product capable of carrying up to 2500 A at voltages rated up to and including 36 kV and a 52 kV

Figure 134
A cold applied
single core joint

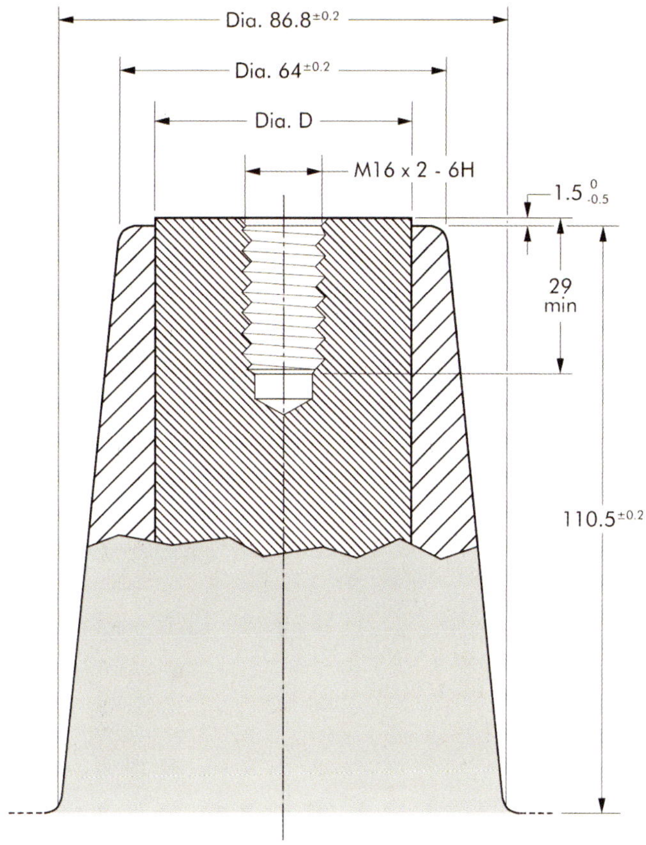

Interface type	Voltage Um (kV)	Current Ir (A)	Dia. D (mm)
F1	36	2500	50
F2	52	630	22 min
F3	52	1250	32

Figure 135
F type bushing

rated product with a current rating of 1250 A. The bushing interface, designated F, is shown in fig. 135.

One other advance has been in the reduction in size of the separable connector whilst still fitting the same bushing profiles.

18.3.2. Reduced size of products

The size of the separable connector has been considerably reduced as shown in fig.136 which compares the original product with the smaller version.

The reduction is achieved using a new design as shown in fig. 137. The projection on the left of the picture removes the need for a connecting plug. The moulded part connected to the bushing is designed to fit standard bushings as usual, but subsequent units have been moulded to fit to the rear of the first and subsequent connectors (fig. 138).

A metal connecting rod allows the extension of the bushing conductor to accommodate additional cables. Figure 138 shows how the connection is made in section.

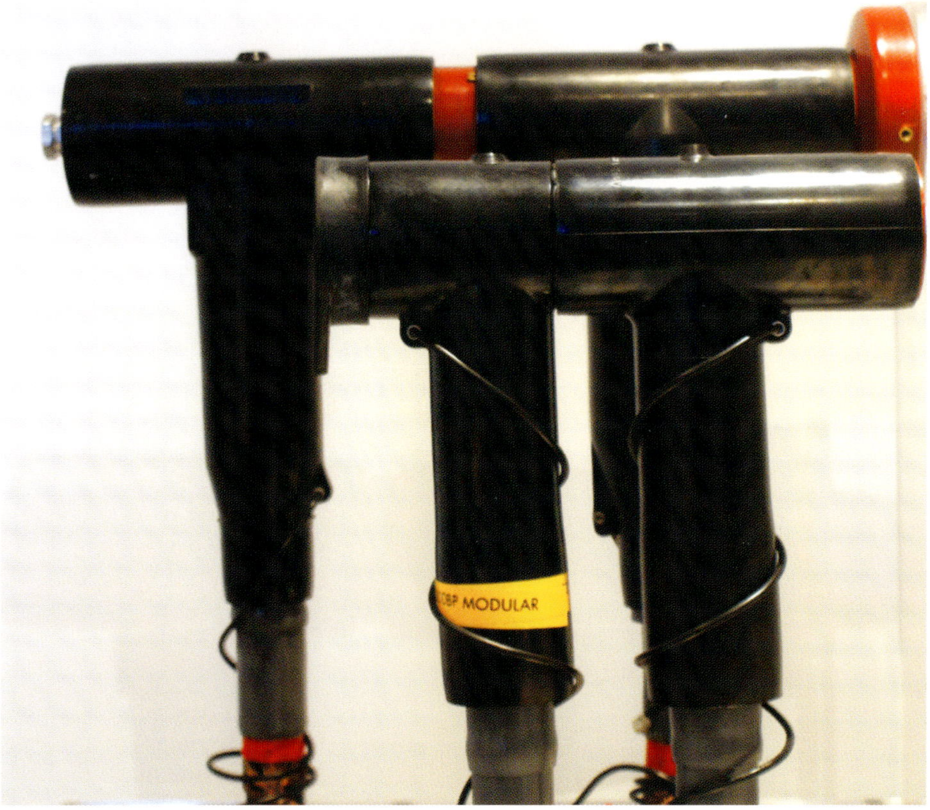

Figure 136
Original connectors to rear new design in front

Figure 137
New moulded connector design

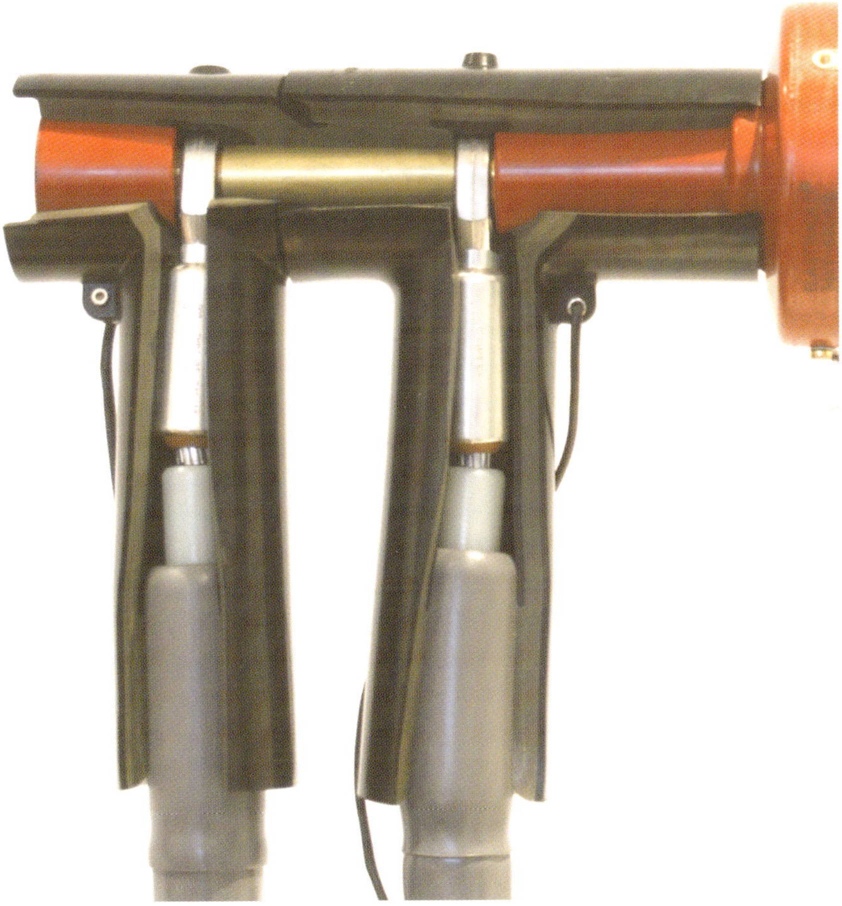

Figure 138
Cut-away of
new piggy back
connections

Chapter 19

A review

19.1. Introduction

Medium voltage cable accessories have evolved over the past fifty years to provide generally reliable systems. There are, unfortunately, some less than perfect products on the market but generally speaking the established manufacturers do not fall into this category. The correct working of accessories is dependent upon many issues from the correct specification of the product through to competent installation. The following is a list of do's and don'ts which should be adhered to.

19.2. Do's and Don'ts

The following is perhaps a list of the obvious, but is worth repeating:

- Do specify and receive the correct product for the application.
- Do check that the cable dimensions are compatible with the accessory. Modern cables using higher conductivity aluminium conductors are still specified in mm^2, but may have a reduced diameter. Furthermore, the cable insulation has been reduced by some manufacturers as material quality and manufacturing processes have improved.
- Do not mix and match products unless it is the only solution in an emergency. Replace with the correct product as quickly as possible.
- Do employ only trained competent staff to install the accessory.
- Do carry out the work in a clean and dry environment if at all possible.
- If in doubt ask – your supervisor or the manufacturer.
- Do use correct, well maintained tools for the job.

- Do read instructions carefully.
- Do prepare the cable carefully and correctly to the manufacturers' dimensions.
- Do not rush any part of the job.
- Do not take short cuts.
- Do make sure that lugs and connectors are correctly installed. For crimp connectors make sure you use the correct dies.
- Do smooth any sharp edges on lugs and connectors caused by crimping dies.
- Do not forget screen cut treatment as specified by the manufacturer.
- Do not forget to apply sealant tapes if specified.
- If components are missing at any stage, stop work and obtain the missing components. You may hide an omission at installation, but sooner or later it will reveal itself – probably with total failure.
- Do not forget to install the earth system correctly.
- Do make sure all components are correctly installed i.e. fully shrunk and correctly positioned.

Note:

With heat shrink systems, heat evenly and gently with a soft flame until the component is completely shrunk. Shrink in the direction shown in the instructions. Do not use a fierce flame as this can lead to uneven shrinking and scorching of components (scorching is not usually detrimental to the working of the accessory but usually indicates poor jointing practice).

With cold shrink systems it may be that if a joint or termination has been stored for several years in an extreme environment, the components may not fully recover to their designed dimensions. This could present a particular problem at the bottom end of the range scale, but in practice it is not likely to be a problem provided the product has not been in storage for more than 2 years. A more likely source of insufficient recovery is incorrect selection of accessory for a given cable. It will be obvious after installation if the components have fully recovered onto the smallest diameter of the joint –if they have not, stop work and seek advice.

- Do check when installing terminations that if a roll or cross is required that it can be performed to meet the clearance required between cores or that if necessary a core-crossing kit is used.
- Do not allow sheds on adjacent cores of a three core termination to touch as this can lead to discharging.

- Do securely connect termination lugs to equipment. A loose connection will cause overheating and failure. This is particularly important on indoor terminations where it can lead to extensive equipment damage.

- Do take care when back filling a joint bay. Polymeric joints are more susceptible to mechanical damage than resin filled systems.

- Do check during pressure testing for any unusual signs such as excessive audible discharging. This is somewhat subjective and the test may well not indicate any abnormality present.

- Do not ignore audible discharge after a termination has been installed. There is a reason for this and it may be the prelude to a problem. A visual inspection may reveal the cause (cores touching, excessive moisture in the cable box etc.). It is better to sort out the problem before breakdown occurs – it is cheaper in the long run. If the problem is not obvious, specialist discharge detection equipment is available to ascertain the problem area.

19.3. The future

The end user of a cable accessory requires that the product he buys and installs today will still be performing satisfactorily at the end of the life of the system. The user is often justifiably reluctant to try new, untried systems and prefers the tried and tested product. As a result of this, it is likely that cable accessories will continue to be much the same as they have been for the past thirty or forty years, with only the delivery system being improved. Factors such as price, play an increasing role in the modern world of power distribution as in all other areas of life and this dictates that manufacturers are constantly striving to reduce manufacturing costs, whilst at the same time maintaining quality and introducing innovation. The user should bear in mind, however, that a few dollars saved today could cost several thousand in a couple of years time.

It is fair to say that since the first dry cable accessories were introduced, little fundamental change has taken place in the design. The joint or termination still requires stress control, insulation and earthing and will continue to do so.

The one inherent potential weakness of joints and terminations is that they are not assembled in a factory and there is often little control over the person installing the product in the field. The trend is for one piece products but this is only half the story. The major problem encountered today is poor cable preparation and unfortunately the accessory manufacturer can only partly cater for this by assisting in giving regular training to jointers. This problem will probably not improve, especially when jointing new cable to old three-core paper cable, because the jointers have less and less practice working with the older cables and their skills deteriorate.

Much of the authors' work in the field of cable accessories has been such that only problems required his attention. It is sometimes easy to forget in such a position, that not every accessory produced finishes up in the failure analysis depot. In general the number of joint and termination failures is very small, but if each reader of this book is able to prevent at least one failure as a result of the knowledge gained, it will have been a worthwhile exercise.

About the Author

Derek Goulsbra began his working life in 1960 as an engineering apprentice with A Reyrolle & Co Ltd., a switchgear manufacturer in the North East of England. He remained with Reyrolle for the next 13 years during which time he was sponsored in his higher education and worked in high voltage insulation laboratories. In 1967 he obtained a B.Sc. in Electrical Engineering after attending a four year sandwich course at Sunderland Technical College. In 1968 he was awarded a M.Sc. in High Voltage Insulation Engineering from the University of Manchester Institute of Science and Technology and after returning to Reyrolle for two years then embarked on a Ph.D. at UMIST. He received his doctorate in 1972 for a thesis entitled "The behaviour of electrical discharges on polluted insulation".

There then followed four years as assistant professor of electrical engineering at the University of Petroleum and Minerals in Saudi Arabia where he worked closely with industry to obtain practical training for students.

On return from Saudi Arabia he worked for three years at GEC Distribution Switchgear in charge of the insulation group and high voltage laboratory.

In 1980 he joined Sigmaform to set up the Middle East Division and continued in this position until 1989 when he became manager of the Technical Centre responsible for all technical matters relating to heat shrink products worldwide. In 1994 the role was extended to lead a project developing cold shrink accessories and this was continued when he moved to Raychem in 1996.

He left Raychem in 2000 to become a consultant.

The author has written numerous papers on cable accessories and presented them in many countries.